RETHINKING the FEDERAL LANDS

RETHINKING the FEDERAL LANDS

Sterling Brubaker, Editor

Resources for the Future, Inc., Washington, D.C.

Published by Resources for the Future, Inc., 1755 Massachusetts Avenue, N.W.,
Washington, D.C. 20036
Resources for the Future books are distributed worldwide by The Johns Hopkins
University Press.

Library of Congress Cataloging in Publication Data
Main entry under title:

Rethinking federal lands.

  Bibliography: p.
  Includes index.
  1. United States—Public lands—Congresses.   2. Land tenure—United States—Con-
gresses.   3. Land use—Government policy—United States—Congresses.   I. Brubaker,
Sterling.   II. Resources for the Future.
HD216.R48   1984   333.1′0973   83–43261
ISBN 0–915707–00–4
ISBN 0–915707–01–2 (pbk.)

Resources for the Future is a nonprofit organization for research and education in the development, conservation, and use of natural resources, including the quality of the environment. It was established in 1952 with the cooperation of the Ford Foundation. Grants for research are accepted from government and private sources only on the condition that RFF shall be solely responsible for the conduct of the research and free to make its results available to the public. Most of the work of Resources for the Future is carried out by its resident staff; part is supported by grants to universities and other nonprofit organizations. Unless otherwise stated, interpretations and conclusions in RFF publications are those of the authors; the organization takes responsibility for the selection of significant subjects for study, the competence of the researchers, and their freedom of inquiry.

This book is a product of RFF's Renewable Resources Division, Kenneth F. Frederick, director. Sterling Brubaker is associate director of the division and a senior fellow. The conference on which this book is based was sponsored by RFF, with funding provided by the A. W. Mellon Foundation and the Lincoln Land Institute.

The book was edited by Jo Hinkel and designed by Elsa B. Williams. The index was prepared by Lorraine and Mark Anderson.

Contents

Foreword   ix
Acknowledgments   xi
List of Contributors   xiii

PART I: INTRODUCTION   xv

1. Issues and Summary of Federal Land Tenure
    Sterling Brubaker   1

PART II: THE FEDERAL LANDS: WHY WE KEPT THEM—
HOW WE USE THEM   33

2. The Federal Lands—Why We Retained Them
    Paul W. Gates   35

3. Why Have We Retained the Federal Lands? An Alternative
    Hypothesis   *Barney Dowdle*   61

4. The Federal Lands Today—Uses and Limits
    Perry R. Hagenstein   74

5. Uses and Limits of the Federal Lands Today—Who Cares
    and How Should the Current Law Work?
    D. Michael Harvey   108

PART III: RETENTION, DISPOSAL, AND PUBLIC
            INTEREST   123

      6.   The Claim for Retention of the Public Lands
           Joseph L. Sax   125

      7.   Weaknesses in the Case for Retention
           Richard L. Stroup   149

      8.   The Case for Divestiture   *B. Delworth Gardner*   156

      9.   Ownership and Outcome—An Economic Analysis of the
           Privatization of Land Tenure on Forest and Rangeland
           Gordon C. Bjork   181

PART IV: INTERMEDIATE POSITIONS AND SPECIAL
           PROBLEMS   193

     10.   Major Alternatives for the Future Management of the Fed-
           eral Lands   *Marion Clawson*   195

     11.   Sharing Federal Multiple-Use Lands—Historic Lessons and
           Speculations for the Future   *John D. Leshy*   235

     12.   Ideology and Public Land Policy—The Current Crisis
           Robert H. Nelson   275

           Index   299

Foreword

The Reagan administration took office in 1981 with what they perceived to be a mandate for curbing the role of the federal government and stimulating growth in the private sector. While virtually all their efforts to reduce the size and influence of the federal establishment have encountered major resistance, few of the administration's proposals have generated more criticism than those that would alter the ownership and management of the federal lands. Proposals to increase energy and mineral development on federal lands, to accelerate timber harvesting in national forests, and especially to expand the sale of federal lands have generated strong and vocal opposition. The ensuing debate has been heated but seldom enlightening as to the relative merits of specific proposals. The discussion has been dominated by ideological views as to the proper role of government versus private enterprise in U.S. society. Evidence offered in support of various positions tends to be fragmentary and anecdotal.

Early in 1981 Resources for the Future undertook two activities designed to examine how ownership status affects the management of federal lands and the total benefits society derives from them. Marion Clawson, who has been active as either an analyst or administrator of federal lands for nearly fifty years, undertook a reconsideration of the federal lands. His book, *The Federal Lands Revisited*, which was funded in part by the Richard King Mellon and the Weyerhaeuser Company foundations and published by Resources for the Future in the fall of

1983, suggests some alternatives to existing tenure and management practices that should satisfy many of the objections of both those who are critical of the existing system and those who fear the implications of large-scale transfers of public lands into private hands.

Simultaneously, Sterling Brubaker started planning for a national workshop on the federal lands to highlight the changing character of federal lands management issues and to examine whether modifications in tenure and management arrangements might better serve the national interest. The workshop did not seek to arrive at conclusions or recommendations. Rather its objective was to enlighten and stimulate new ideas by providing a forum where people of widely divergent views and backgrounds could present and discuss their positions in a professional manner. The workshop, held in Portland, Oregon, in September 1982, certainly provided the audience with a better understanding of the importance of the federal lands, the advantages and disadvantages of the present as well as alternative tenure and management arrangements, and the concerns of all parties.

Rethinking the Federal Lands, edited by Sterling Brubaker, draws largely on the papers and discussion of the Portland workshop. This collection of papers and Brubaker's excellent introduction and summary enable a broad audience to obtain a better understanding of the issues surrounding what in recent years has become one of the nation's most contentious resource questions.

Washington, D.C.   Kenneth D. Frederick,
September 1983   Director, Renewable
                                                    Resources Division,
                                                  Resources for the Future

Acknowledgments

Most of the papers in this volume grew out of an RFF conference on the subject of tenure arrangements on the federal lands. That conference in turn evolved from extensive consultation with members of the RFF staff and with an advisory group assembled to discuss the issues and the people knowledgeable about them.

Members of the RFF staff who were most helpful in the initial discussions include Emery N. Castle, Kenneth D. Frederick, Allen V. Kneese, John V. Krutilla, Hans H. Landsberg, and W. David Montgomery. Subsequently Christopher K. Leman, Robert Cameron Mitchell, and Roger A. Sedjo of the RFF staff provided valuable advice on who should be invited to participate in the discussion. Throughout, Marion Clawson, who also is a major contributor to this volume, offered useful perspectives both on issues and people.

The advisory group included Jack Artz, then at the U.S. Department of Agriculture, Ralph Hodges of the National Forest Products Association, Cynthia E. Huston of the Congressional Research Service, Robert H. Nelson of the U.S. Department of the Interior, William Northdurft, private consultant, and Ross Whaley of the U.S. Forest Service. Many other individuals and institutions were solicited for comment and their total contribution was very helpful.

At all stages of the process, Anthony T. Stout, of the RFF research staff, was an energetic participant, both in making arrangements for the conference and in contributing ideas and criticism as it unfolded; his assistance has been invaluable. Jo Hinkel, of the RFF editorial staff, labored hard to convert the contributions of diverse authors into a reasonably consistent and readable document; only the volume editor can fully appreciate how much she has done. Maybelle Frashure handled the heavy load of conference correspondence and manuscript typing most competently.

Finally, special thanks go to the A. W. Mellon Foundation, as well as to the Lincoln Institute of Land Policy, who provided financial support for this endeavor.

As editor I cannot claim all of the responsibility for criticism that may be directed at the ideas expressed here. Each contributing author has been free to present his own views in his own language, and to them must go most of the credit and some of the blame. However, I do assume responsibility for initiating an RFF discussion of this topic and for the overall structure and balance of the volume. No investigation of this topic can be complete or definitive, but this one is intended as a serious, open-minded discussion of an issue that is important to us all. If it falls short of that, then I am at fault.

                                                                    S.B.

List of Contributors

Gordon C. Bjork is Lovelace professor of economics at Claremont McKenna College and Claremont Graduate School, Claremont, California.

Sterling Brubaker is associate director and senior fellow in the Renewable Resources Division of Resources for the Future, Washington, D.C.

Marion Clawson is senior fellow emeritus in the Renewable Resources Division of Resources for the Future, Washington, D.C.

Barney Dowdle is professor of forestry and adjunct professor of economics at the University of Washington, Seattle.

Paul W. Gates is emeritus professor of American History at Cornell University, Ithaca, New York.

B. Delworth Gardner is professor of agricultural economics at the University of California, Davis.

Perry R. Hagenstein is president of Resource Issues, Inc., Wayland, Massachusetts.

D. Michael Harvey is chief counsel for the minority on the U.S. Senate Committee on Energy and Natural Resources, Washington, D.C.

John D. Leshy is professor of law in the College of Law at Arizona State University, Tempe.

Robert H. Nelson is a member of the economics staff in the Office of Policy Analysis of the U.S. Department of the Interior, Washington, D.C.

Joseph L. Sax is Philip A. Hart university professor at the University of Michigan Law School, Ann Arbor.

Richard L. Stroup is director of the Office of Policy Analysis of the U.S. Department of the Interior, Washington, D.C., and professor of economics at Montana State University, Bozeman.

I
Introduction

1

Issues and Summary of Federal Land Tenure

STERLING BRUBAKER

For several decades the basic tenure arrangements for the federal lands have been little questioned, but lately that quiescence has been disturbed. Although apparently only a short-blooming desert flower, the Sagebrush Rebellion attracted attention to the tenure question. The Reagan administration has sustained it, to a degree, with its proposals for disposal of some lands, although the administration has avoided placing the fundamental issue high on the policy agenda. A more far-reaching challenge to federal land policy has come from a handful of libertarians and their academic allies. Aside from these recent stirrings, however, the question of tenure has not been reevaluated in public or academic debate for many years.

Resources for the Future, in September 1982, organized a national workshop to give serious and balanced attention to this question, focusing mainly on those resource lands administered by the Forest Service and the Bureau of Land Management. The present volume derives largely from the discussion at that workshop, which was held in Portland, Oregon.

The federal government is by far the largest landowner in the United States. A major body of law and a sizable bureaucracy govern the management of that land. Controversies over federal lands abound, but for a number of reasons, management issues have dominated public discussion to the exclusion of basic questions of tenure.

1

On the face of it, it is somewhat of an anomaly for the federal government to hold vast acreages of land in an economy where the prevailing ideology favors private ownership. In other sectors of the economy there is a marked bias against the government holding productive assets not used directly in government business. To be sure, there are a few conspicuous exceptions (for example, federal power dams), but most government acquisitions do not remain in public hands for years without challenge. What is the rationale for the different attitude with regard to the federal lands?

Note that during the present century the trend clearly has been toward retention rather than disposal of federal land. Federal ownership of timberlands has been substantially without challenge since the Forest Service Organic Act was passed in 1897. The principle was reaffirmed and expanded by the Weeks Act in 1911 and by much subsequent legislation. Grazing lands remained open to private claimants until 1934, when they were effectively closed upon passage of the Taylor Grazing Act. Retention of forest and grazing lands was further confirmed by the 1971 report of the Public Land Law Review Commission (PLLRC) and by the Federal Land Policy and Management Act (FLPMA) in 1976. Many would argue, as Michael Harvey does in chapter 5, that the question of tenure is a settled issue and need not be reopened. However, what one Congress does, another may undo. Even if, politically, the current prospects for change are slight, forward thinking can elevate and influence debate should the pendulum swing.

One way in which the pendulum has swung is toward greater skepticism over the effectiveness of government in solving many types of problems. Earlier legislation that had the effect of retaining federal lands reflected a prevailing attitude of suspicion toward private business and a faith in governmental solutions. This arose out of the progressive movement of Theodore Roosevelt and the long heritage of Franklin Roosevelt's New Deal. Few have the same confidence today in the role of government, and the Reagan administration articulates a desire on the part of a significant portion of the public to limit the scale and reach of government.

The original rationale for retaining land in federal ownership is traced by Paul Gates and further developed by Robert Nelson in this volume. It stemmed from public reaction against what was seen as "cut and run" exploitation of forestland that, it was feared, would leave the nation short of timber. By contrast, government was expected to take the long view that private firms could not. The resource would be managed according to scientific principles for sustained yield. Subsequent additions to forest reserves in the eastern United States provided a home for stripped, abandoned, and often tax-delinquent land that was contributing to watershed and flood control problems. The grazing lands that were withdrawn in 1934 were those that had not been claimed by settlers. Government retention and management was a way of regulating access to land that private owners could not afford to acquire and for which states were reluctant to assume responsibility. Again, progress toward scientific management was thought to be more compatible with government ownership.

The original rationale for retaining federal resource lands has been much attenuated by now. Expertise in forestry and in rangeland management is no longer an exclusive preserve of the federal government—private firms and state governments also have competent staff. As forestry moves from harvesting old growth to timber cropping, private motivation to provide for sustained yield is strong. Private graziers are in like position, powerfully motivated by the market to preserve their lands. Land is more valuable today, and very much less federal land would go unclaimed now than in the past; it would be sought, whether for commodity production or recreational use.

Perhaps new rationales have arisen to replace the original arguments for retention. If so, they need to be more fully articulated. The current stress on multiple use and on environmental and wilderness protection does emphasize values found on some lands that might be neglected were the lands in private ownership. Perhaps this rationale can be extended to the most ordinary federal lands (see Sax in chapter 6), but that is a matter for argument. In any case, those who currently favor retention of ordinary land must put a considerable part of the burden on the argument that public ownership protects or enhances values that are neglected privately.

Many knowledgeable people who are concerned with the federal lands recognize that federal management faces severe problems. Some would focus all attention on the solution of those problems, seeing talk of disposal as a diversion. They deny that there is a problem calling for so drastic a remedy as disposal. Yet many others have noted that recent attempts to rationalize federal planning on resource lands have led to current costs far exceeding federal revenues from surface uses, even though no charge is made for the capital value of the land. At the same time, it is alleged that more intensive federal management may have led to exploitation and development of some land that private entrepreneurs would have left idle. Perhaps managerial improvements could reverse these results, but it is fair to ask whether the results are endemic and unavoidable under federal management in today's policy world.

Issues and Ideologies

The question of retention or disposal of federal land is first of all an ideological issue. The prevailing ideology in the United States favors private ownership of the means of production, and virtually all of the public rhetoric bows in that direction. Many partisans of enterprise demand that reality conform with that ideology. The acceptance of that position is such that public landholdings are almost never defended in terms of an overtly socialist ideology. Yet many sophisticated people harbor private doubts about the morality and results of a market enterprise system that may appear in conflict with their religious and humanistic values, while even more of the less privileged in the society lack any ardent allegiance to that system. Thus the ideological conflict is present, though muted, with those favoring disposal usually openly avowing their ideological position, while their opponents camouflage theirs with other arguments. Of course, neither the proponents of disposal nor the advocates of retention rest their case on purely ideological arguments.

The basic case for the market system is unchanged from the time of classic liberalism. It holds that individuals best know their own interest and that, when they are free to seek it through exchange in markets that are responsive, the resulting output will be at the social optimum. Markets work best where property rights in land, goods, and labor are clear and predictable and can be freely exchanged.

All admit that markets may not be fully responsive for reasons of monopoly or other imperfections, but they may differ on the significance they attribute to that failing. Likewise, all admit that the outcome is shaped by the existing distribution of wealth and income so that the optimizing that occurs, while appropriate to that distribution, may not meet independent standards of equity.

Critics note that some goods (and some bads) fall outside the usual reach of the market. Environmental damages are a common example, since the costs that they entail frequently are not borne by those who generate them, but rather are imposed on others. Theoretically, this defect can be cured if such costs and benefits can be brought within the calculations of economic actors (internalized). In practice it often is very difficult to extend the system of rights and markets in this way. Moreover, where the effects of actions taken currently are to be felt by remote generations, it at least is debatable whether the market or the political process registers their interest better.

Belief in the market involves commitment to a metaphysical system, as Gordon Bjork phrases it in chapter 9. It appeals intellectually because of its internal consistency once the basic premises are accepted and once the responsibility for its equity outcome is assigned elsewhere. It offers a standard for optimization, avoiding the arbitrariness of most other approaches. Another major appeal of the market system to those who concede its imperfections is the lack of preferred substitutes. Alternative approaches appear to depend too much on goodwill, good judgment, or improbably good institutions.

Of course, markets can still operate as allocative devices under constrained conditions, even if the outcome is nonoptimal relative to the market ideal. Those who are unwilling to support the metaphysical system will focus on how to constrain the system to yield the results that they desire. Inevitably, such constraints are imposed through the political process and they may or may not make use of market devices. However, if the process is not democratic or if it allows excessive influence to narrow interest groups, the outcome may appear arbitrary as well as inefficient. The definition of the appropriate constituency for making the political decisions governing resources also is far from a simple matter.

Admitting that there are economic values not generally well served by the usual operation of atomistic markets, does this imply resorting to governmental action to secure such values? Not necessarily, is the response of those committed to market solutions. They would extend the reach of markets by creating marketable rights to presently nonmarketable values so as to bring market incentives into the service of such values. The existence of wilderness is recognized as a value that must be weighed against the usual commercial value of development of the same land. If property rights to wilderness land were acquired by those dedicated to preservation, then, it is argued, they would face the issue of limiting exploitation of resources from the land in a more responsible way, since returns from controlled exploitation might be employed to protect even more land elsewhere. Without such rights, wilderness advocates are tempted to use the political process to exclude reasonable uses because they face no opportunity cost of so doing. As another example, developers in their own interest may choose environmentally sound procedures for development if they stand to capture the gain in value from so doing. Ingenious extensions of markets in this view are believed capable of incorporating most externalities, leaving very few public goods that would require public action to produce at the optimum level.

Adherents of this position usually minimize the importance of true public goods or externalities because they feel the values of concern mostly can be expressed through markets once clear rights are established. They also take a very jaundiced view of government action. At the heart of this view is the alleged absence of any real responsibility in government. Pressure groups that influence government policy do not themselves bear the social costs of their actions. Yet they prevail because the general public is ill-informed, disinterested, has very diffused interests, or is lacking in effective channels for monitoring and repelling the actions of the few. The bureaucracy also is seen as playing a game of its own, having institutional interests that may not promote optimum solutions, and operating in a climate of ambiguity where clear standards of performance are absent. Thus, an empire-building Forest Service bureaucracy can be charged with subsidizing the harvest of slow-growing Rocky Mountain forests that private firms would find uneconomical, thereby defeating both preservationist values and economic rationality.

The market system relies on the benign overall results of everyone pursuing his or her individual interest. Being thereby motivated to be efficient maximizers, or else unsuccessful, individuals innovate, strive, master, and produce. They also may cooperate, but they can do so freely as individuals pursuing private interests who enter into covenants with other like-motivated individuals in pursuit of common objectives.

But not all behavior fits this pattern, as Joseph Sax notes. People also form collectives and surrender some of their autonomy to them. By the fact of belonging, they allow their own values to be shaped in ways different from the purely autonomous individual. Note that the market does not inquire into how preferences are formed, taking them as given, and makes no judgment of their value.

The evidence for collective behavior is found in countless voluntary organizations among which churches are perhaps the most fascinating because the individual often allows the collective to form his views. But in other cases, as well, the collectivity not only reflects but also may influence individual preferences by controlling the flow of information and providing an ambience in which a common view is reinforced. This can be quite pervasive, for in some sense all individual preferences are formed in a social process, and individuals always can choose to express their preferences in collective rather than individual form. Some benefits are so widespread (common defense or public streets, for example) that it may be impossible to provide them only to those individuals who are prepared to express a market preference for them; in such cases collective action is widely accepted.

The government usually provides those public goods with broadest appeal. There are inherent problems in deciding how much to provide. What is the efficient level of production? If market surrogates are sought to answer this question, then a formal kind of criterion can be introduced, but since individuals do not actually pay the costs imputed to them, this is of doubtful value. Collective action may be the only effective way to secure some values. It even may be more efficient in some range of actions, but usually it lacks clear efficiency criteria. Because the state is an involuntary collective and has coercive powers, contrary views are not registered as easily as in markets.

In practice, decisions about the disposition of federal lands have always had a large element of social philosophy in which the market model never has prevailed fully. Thomas Jefferson saw the provision of land to independent yeomen farmers as the basis of a stable democratic system. It was that political stability that attracted him, not just the ideal of economic efficiency or apolitical individualism. Subsequent disputes about how land should be disposed of reflected basic views of society—whether free or slave, whether smallholder or large landowner.

Later, when private operators were perceived as destructive and short-sighted, government retention of forest and grazing lands was defended as permitting rational management and wider distribution of benefits. Curiously, this view was advanced long before the era of intensive management had arrived. Growing out of the New Deal years was an attitude that government had an active and direct role to play and that the market had shown itself incapable of handling many problems of production, to say nothing of conservation and equity. Decisions would be made politically. So long as the number of actors and interests remained comparatively small, political decisions seemed manageable. With the proliferation of environmental-preservationist concerns in the past two decades crowding in against the already conflicted commodity uses of land, the political decision process has become very complex and the results appear to have gone adrift. To those advocates of political solutions, it almost appears as though any political outcome is considered equally appropriate so long as the process allows for the various interests to register the direction and intensity of their views. There is no independent criterion of the public interest.

To summarize, while the ideological argument in favor of disposal seldom is confronted directly, its force is blunted by those who point to the various failings of the market, by positive arguments for collective choice in some areas, and by the embedded interests accustomed to battling for their views in the political arena. As a result, the federal land system has great inertia, as all observers have noted, and even the most marginal changes become the occasion for emotional battles with ideological overtones.

Asset Management and the Federal Lands

Apart from ideological questions, does it make sense for the federal government to continue to hold major assets in the form of land? This question is much narrower than whether a social optimum is achieved by holding or disposing. Political and equity issues, as well as the relative efficiency with which the land is employed in different tenures, all are part of the larger question of the social optimum. But from an asset-management point of view the government is no different from any other property owner; it should hold that form of asset that maximizes its position.

It the government wishes to maximize its asset position, it would need to consider what is happening to the value of its assets compared to the carrying charges on an equal volume of debt. One cannot know what the future will hold in this respect, but Marion Clawson's estimates of the growth in value of the federal lands (see chapter 10) suggest that they have not been a bad speculation, rising 22-fold in value since the 1920s. Federal lands also yield revenues and incur operating charges; these must be entered into the balance. Considering all these factors, in all circumstances, one could not exclude as inappropriate an asset-management argument for disposal of federal lands, but the case has not been convincingly established. If asset management is used as a criterion, then all types of assets, not just federal resource lands, should properly be considered for disposal or retention. Of course, none of the forgoing suggests that asset-management arguments should necessarily be controlling.

We need not consider the purely expedient argument for selling land to redress current budgetary deficits. Few would endorse that idea in naked form. An administration with large deficits may seek to divert attention by asserting that sales will be used to reduce the national debt, but this is unconvincing; it offers the spectacle of the government paying off debts with one hand while contracting more with the other.

In the absence of current deficits, is there a rationale for using assets to reduce the debt? That is not obviously the case. The government faces no threat of insolvency through inability to service a stable debt, and such a debt is no burden on the private economy.

Government debt instruments play a major role in financial markets, and diminishing their supply would have disruptive and not entirely predictable results. Certainly any rapid disposal of federal lands would be difficult to accomplish without massive disruption of both land and financial markets. Land prices in the West, both public and private, probably would be forced down, and private financial assets now held as bonds would be converted from highly liquid to highly illiquid form.

Who could be confident that such large-scale transfers could be accomplished without great inequity and fraud—a cost to the social fabric that cannot be ignored. Note that those who are most skeptical about the capacity of the bureaucracy to manage would still depend on it to conduct an effective sale.

While the asset management case for large-scale federal land disposal may be shaky, the Reagan administration and some in Congress have proposed more modest and specific disposals that can be considered on narrower grounds. It long has been federal policy to dispose of surplus federal real property, and all serious students of the federal lands admit that many federal holdings cannot be managed efficiently because of their size, location, or character.

Over a period of decades, up to 5 million acres of BLM lands have been identified as ill-suited to federal management for one reason or another. Such land is now being made available for disposal. As much as 5 percent of federal land, or 35 million acres, have been suggested as the limit to what might be disposed of under the administration program. Note that sales of this magnitude would have a negligible effect on the public debt. While a change in the status of any specific tract of land is apt to arouse argument, few would argue about the wisdom of selling idle land overtaken by urban development or of consolidating dispersed management units. Up to now the emphasis has been more on exchange than sale, but exchange has proved a very unwieldy procedure. In the final analysis, resistance to the official asset-management policy of disposal seems to stem more from an ideological opposition to any reduction in the federal estate than to the specifics of what is being proposed.

Some outside the administration have embraced the more expansive view of the asset-management case for disposal, arguing that the government would be well-advised on those grounds to sell most of the federal lands. However, it remains unclear why the government would be better off with less of both land and debt. If such a case is made, it might derive from the high cost to the government of managing federal lands. As Robert Nelson and Gabriel Joseph have reported, management of the federal surface estate on forest and grazing lands costs several times the revenues from those uses.[1] Thus a continuing budgetary drain could be averted by disposal. The existence of a current operating deficit acts to offset possible capital gains that may accrue from holding and thereby weakens the argument for holding.

Economic Efficiency and the Public Lands

The most substantial arguments for disposal of federal lands hinge on
the assumption of the greater efficiency of private management. This is
an article of faith on the part of all of those inclined to disposal. Curiously, as Richard Nelson has pointed out, the superiority of private
management is not documented in detailed empirical studies.[2] The evidence offered for it is often anecdotal, grounded in comparative economic systems, or found in logical deductions about behavior from prior
acceptance of the market as a metaphysical system. It may remain a
plausible proposition, but as such it is less compelling than an established
fact.

As Richard Nelson observes, the case for private efficiency is not
based on welfare economics. He reminds us that the theorems of welfare
economics—namely, the existence of a competitive equilibrium and of
a social optimum—often are not met in practice. Information is imperfect, markets are not always atomistic, and externalities abound. Moreover, the outcome is conditioned by the prior distribution of rights, and
any mechanism for intergenerational equity is imperfect. Some of those
who agree with this critique, however, still might argue that the system
tends in the direction of optimization.

The efficiency argument is more convincingly made on microeconomic
grounds that private firms, under the discipline of the market, are motivated to operate a responsive and innovative enterprise in the most
economical way so as to produce a socially better outcome than would
result from alternative systems. This commonly held position is the main
economic rationale for private enterprise. To it may be added a philosophical conviction that personal liberty can thrive only when the coercive power of the state is limited in the economic sphere.

This is not the place to argue the philosophical position, although it
should be noted that personal liberties quite like our own seem to have
survived in some countries with significantly larger public sectors. Likewise, the evidence from comparative systems is not entirely compelling.
While centrally planned socialist economies have not fared very well,
neither have all enterprise economies; the mixed, highly interventionist
economies of northern Europe and Japan have as good a growth record
in postwar years as any. Nonetheless, the motivation and flexibility of
private management, especially on the investment side, seems to give
it a dynamic that is hard for public managers to match. Lamentably, it
is hard to point to studies that systematically examine the question.

Those who remain skeptical of the case for enterprise would note that
large firms are staffed by private bureaucrats who behave much like

their public brethern. They would hold that a well-administered public bureaucracy can manage as well as a private one. Although bureaucratic motivations differ, there have been examples of effective performance. Recall that the original rationale for public administration of resources was an appeal to scientific management.

For the most part, however, the case for public management currently does not rest on any presumption of the superiority of public management; rather it only assumes that public managers are equally as competent as private. Instead the argument returns to social efficiency concepts, holding that private operators will not produce the socially optimal quantity and mix of goods. In particular, they will neglect the production of public goods.

If true public goods are insignificant, then the case for public management is weaker. Those opposed to public management argue precisely that the weight of public goods is small. Some take this position, although they would be willing to concede a role for public management in cases where public goods and externalities are important. Others have argued ingeniously that values normally thought of as public goods, such as wilderness or open-space preservation, can be provided efficiently through voluntary collectives or private covenants. However, there are distributional consequences of so doing and the rules of access also may change, so that even a convincing demonstration of this possibility would not meet with universal approval.

Public management often is seen as being more consistent with the rational control of externalities. After all, it does provide for a large decision unit that may make internalization feasible and it lacks the private motive to divert negative consequences elsewhere. However, public bureaucracies also have budget constraints that militate against costly control measures. In general, our control measures have been based on regulation rather than on market incentives, and it is unclear that bureaucracies are any easier to regulate than private firms in this area.

Laws governing public land management provide for many objectives, management criteria, and procedures, not all of them consistent with one another. On their face they do not demand the achievement of a social optimum as defined by welfare economics criteria. It has been argued, however, that the main thrust of court interpretations of the laws has been to give primacy to economic criteria.[3] Since these criteria are developed for a competitive market system, the task of the federal manager under this guidance would become that of seeking an outcome equivalent to what the market would yield were it free to operate in this sphere, with the extension that public managers, not constrained by actual markets, would be free to produce nonmarket goods for which sufficient net values could be imputed.

This point of view has not won wide acceptance. Its main appeal would be an effort to formally incorporate nonmarket goods not provided by private management. However, it would leave the determination of the output mix and level to essentially market criteria rather than political decision. Some public land values now strongly defended politically would be unsupported by economic criteria. Since a combination of subsidies and regulation probably could motivate private managers to produce desired nonmarket goods while still allowing for the presumed efficiency of private management, there seems little need for public management when the aim is to produce a market-defined result.

A recent paper by Krutilla, Fisher, Hyde, and Smith summarizes the arguments for public and private ownership.[4] They conclude that both public and private ownership fail to meet the test of allocative efficiency. They note that private markets do not meet the demanding conditions for efficiency, that they fail to provide optimal public goods, and that they face ownership externalities and lumpy inputs that inhibit desired output. Political choice becomes the only way for unorganized consumers of public goods to express themselves. At the same time, public managers lack unambiguous criteria for efficient management and face incentives and political pressures that do not favor efficiency. Public managers tend to compete in the generosity with which they supply public goods, therefore tilting toward excess. The authors conclude that some of both types of ownership is desirable to provide the type of goods that each does best. While Krutilla and his coauthors would monitor and critique public management, they would not greatly disturb existing ownership patterns.

In chapter 6 Sax argues that the collectivity—in this case, the federal government—may identify values in public lands that differ from the individual values treated by markets. Being true preferences, these values have equal weight with individual preferences. Such values need to be more clearly specified. What do we want from federal lands that differs from private? Vague references to recreation, watershed, and wildlife do not suffice, for these are obtainable on private lands as well and, with appropriate government initiative, could be had in much the same quantity and quality from public lands converted to private ownership. If the reference is to the sheer existence value of wild land, then there appears to be a true public good concerning which collective choices are appropriate, but it is one whose value is very difficult to establish by economic standards. Economic analysis does not appear to help in making these choices except to the extent that it may establish the opportunity cost of the choice. That opportunity cost, while informing the collective decision, need not control it.

If we grant that public goods are of some significance and that not all collective choices concerning them can be made according to economic criteria, then we have a type of rationale for public ownership that lies outside of the usual economic efficiency paradigm for defining a social optimum. Clearly the decision will be made in a political process where procedural rules and democratic process are essential to the validity of the outcome, but there is no definition of efficiency to which that outcome can be compared. This self-validating quality and the absence of external criteria troubles those steeped in the economics tradition; those with a background in politics, law, or political science appear more comfortable with it.

Continuity and Change

The present federal land-management system is highly resistant to change. Few would argue that it is very efficient, but it represents a balance of forces that has evolved over time. The parties at interest, while willing to jockey for advantage, tend to view major changes as risky to them. Federal subsidies to various users lubricate the ill-matched parts and keep the apparatus from grinding to a halt.

Ranchers once were ardent Sagebrush Rebels, but their ardor has cooled with the realization that oil companies and urban interests well might outbid them for the public grazing lands that they now use at below market prices. If they were offered concessionary terms and preferential rights to acquire the land, they might be interested, but absent that, they have become more cautious. Miners are interested mostly in ready access for exploration and exploitation of mineral deposits. They grouse about the withdrawal of wilderness land from mineral entry and the pace of leasing. But they pay nothing for the hardrock minerals that they take from federal lands, and they have succeeded in accelerating lease offerings. Their complaints have an element of bluster, and they have not led any charge for privatization. The forest industry is badly split on this issue. Some firms would welcome a chance to buy productive timberlands, but others are more dependent on the right to bid for federal stumpage. Most recreational use is free or at nominal charge. Recreational users dispute among themselves over the kind of use that should be given preference but do not seek change of tenure. State and local governments in areas with federal lands already receive the lion's share of revenues generated on federal land. They may envy selected parcels and they certainly argue about the allocative formulas that favor some governments more than others, but only a few western government entities formally have sought a change in status of the federal lands. Most sense that they already have a good deal. As John Leshy points out in chapter 11, a change in status also would create real questions about water rights that are so important in much of the West.

Those who lose under the present arrangements are nonusers—primarily eastern taxpayers who subsidize western commodity producers and recreationists. Most easterners are not well-informed on public land issues and do not feel vitally interested. Those easterners who are most concerned often share conservationist and environmental values which lead them to become defenders of the status quo. As was noted earlier, the principal advocates of disposal have been a coterie of academics and conservative think tanks and advocacy groups. Their posture is motivated more by ideology than by their private economic interest.

Yet ideas are important, and ideological changes can give them better reception. Moreover, the attempts of the land-management agencies to introduce more sophisticated management techniques, being very recent, have not yet been shown to be workable. If they do not work, they may point up the excessive planning and management attention given to much federal land and raise anew the demand for change of status.

Ideas for change abound. Clawson develops some of them in chapter 10—all the way from outright sale to retention and improved management. His intermediate positions—especially the suggestions for long-term leasing—are of special interest. Taking a different approach in chapter 12, Robert Nelson observes that much federal land, although designated as multiple use, in truth is suitable mostly for low-density recreation. He would distinguish the high-quality, commodity-producing lands from those suited mostly to recreation and offer the commodity lands to private owners. Recreation lands of national significance would remain in federal hands, but those used mostly by area residents would be offered to state and local authorities for them to manage. Both of these schemes would greatly reduce the amount of land in direct federal management, though a long-term leasing plan does leave formal title intact. However, both also agree (Clawson more than Nelson) that the federal role will remain large, both because some lands properly should remain federal and because other claimants will not be forthcoming for some of the less-desirable acreage.

Leasing appears to be the most politically palatable of the devices for enlarging the role of private management. A truly long term lease differs little from a sale. It gives the operational flexibility sought but escapes the finality of a sale. Conditions can be imposed on land use under either a lease or a sale, but enforcement is likely to be more active in the case of a lease where the government retains an interest in the land. Of course, environmental laws and regulations are equally applicable to leased or private lands, although at the time of a change in status there is an opportunity to impose conditions that will reinforce their effect.

Any shift toward greater private control of the surface estate is sure to be accompanied by changes in access to subsurface resources. Should control of the subsurface be passed with the surface? There is no strong reason why it should. Friction between surface and subsurface owners can be moderated by a system of compensation. Separation of the two property rights is common in the United States, as it is elsewhere where the state commonly owns mineral rights. Since miners have free access to hardrock minerals on public lands, they are likely to resist changes that impair exploration or that compel them to pay full value for deposits. If subsurface rights are allowed to pass with control over the surface before the existence of mineral values is established, then the government gets nothing for its property. While this is already the case for locatable hardrock minerals, and federal returns from onshore leases also are modest, this can hardly be the objective sought by policy. So long as the terms of any change in status are tailored to invite exploration, there is scant reason to pass mineral rights along to those acquiring the surface.

Those who resist a change in the status of federal lands often do so out of concern that it would benefit special interests. Yet special interests, playing what they see as a zero-sum game, already strongly influence the terms of use of the land (see Leshy in chapter 11). Our problem may be that of excessive continuity and the difficulty of change. Even a forceful policy of disposal might require decades, would surely be subject to enervating conditions of disposal, and would leave much land in federal hands. Therefore, some boldness in thinking probably does not carry much hazard of rash action.

The editor's personal reaction to the argument for change can be stated quite briefly. The classic conservationist arguments for the retention of federal commodity-producing lands seem to have lost their force. Private operators now have the incentive and the capacity to provide for the future flow of timber and grazing. Still, the time apparently is not ripe for drastic change.

Those who argue for private management do so largely on the grounds that it is more efficient than public management. They make a good presumptive case, but it has not been documented by the kind of careful comparative studies that normally would be expected to precede a major policy change. Studies of this nature, which necessarily would have to be made by nonadvocacy scholars and institutions, should be supported by both sides of the argument.

At the same time, those who argue for retention do so most effectively by invoking the public goods producible on government-owned land that would be lost if the land were privately held. However, they commonly are vague about the nature of such goods, how important they are, and the optimum amount that should be produced. They also often neglect the fact that public goods may be produced on private land and that a combination of incentives and regulation of private land use might yield the desired amount of public goods without the presumed efficiency penalty that public management imposes on commodity production. A more direct confrontation of these issues is called for from those who would continue to hold the land in public ownership.

The potential for negative externalities resulting from a change in tenure on federal land should not be an issue. If divested, the land should be subject to the same regulatory constraints as in other sectors of the society. (A note of caution here—agricultural land has escaped much regulation to date, both because of the complexity of administering it and because of political resistance.) Such regulation, combined with incentives where necessary, also could be used to reinforce the social interest in maintaining the productive capability of the land should private operations abuse it.

Much federal land is not suited to commercial commodity production and its main value is for recreation. Most such land seems a likely candidate for continued public ownership, whether federal, state, or local. But unless it is of parklike quality, there is no reason to preclude private development of recreational features where commercial operation is feasible.

There is very little political constituency at present for major change in status on federal lands. The most feasible avenue for change probably would be through an extension of long-term leasing along the lines suggested by Clawson. This would allow for realization of the presumed advantages of private management without formal transfer of title, and it could be seen as an evolutionary development rather than an abrupt shift.

In sum, I would favor both more careful study of the evidence for and against retention while at the same time moving in an experimental way toward less federal exclusivity. More active sales and exchanges of dispersed tracts, some testing of long-term leasing for commodity lands, and a more receptive attitude toward state and local acquisition of recreation lands all could be undertaken without final commitment to wholesale divestiture. The experience gained from these initiatives might move the argument away from the purely hypothetical and ideological into the more familiar terrain of American pragmatism.

Summary

Most of the papers in this volume were presented at a two-day workshop that sought to involve knowledgeable persons from the federal land-management agencies, from state and local governments, from affected user groups and businesses, and from academia. The workshop concluded with a symposium at which spokesmen for selected groups with an interest in federal lands were invited to respond to the entire discussion.

The structure of the meeting is reflected in the chapters that follow, with papers and discussion papers produced as separate chapters. Following this introduction, Part I provides background in which the history of federal lands is reviewed and their current use and status detailed. Conceptual arguments over disposal and retention are considered in Part II. Part III examines intermediate possibilities between full retention and disposal and some of the special problems for water, minerals, recreation, and the like that would be created by any change of status. The meeting concluded with the aforementioned symposium, but as formal papers were not requested for that, the discussion is only briefly summarized here.

The question faced by Paul Gates in his review of the history of the federal lands was, Why, after a century of aggressive land disposal, did policy shift toward retention of the land? In chapter 2 he gives the resounding answer that it was done in the national interest and was mostly a response to shortsighted and reckless use of the land by private operators. The shift was mediated by the powerful personalities of those reformers whose names are associated with the conservation movement, but they in turn were the product of their era. At a time when the broader social movement of progressivism was attacking monopoly and corruption, conservationists were able to play on these sentiments and on fears of a timber shortage to advance their cause. While some of the reform movement was tinged with socialism, its main thrust was toward scientific management of resources and concern for the longer-range future. Moreover, as Gates reminds us, some degraded and often tax-delinquent eastern forest and watershed lands were virtually abandoned, inviting a public role to control erosion and flooding. An offer by the

Hoover administration to cede western rangelands to the states found no takers because of the low value and cost of administering the land.

Barney Dowdle, while not disputing Gates's record of how it came to pass, questions the original rationale advanced to justify the shift to retention. Dowdle, in chapter 3, argues that the nation originally had such large inventories of timber that it made no economic sense for anyone to grow trees. In that situation, inventory reduction (read "cut and run") was rational, and it is only in recent years that tree farming made sense. The private sector has recognized this and now engages in efficient forest practices. Much of the concern about timber practices and fear of shortage was based on lack of appreciation for these economic facts, in his view. What was decried as corruption in land disposal is seen by Dowdle as the largely inevitable cost of the socially desirable objective of moving land rapidly from the public to the private sector where it could be more productive.

Commenting on the history of this decision in chapter 12, Robert Nelson stresses the role of the prevailing intellectual climate as a factor in making the shift. Progressivism was committed to scientific management but feared monopoly. It saw government ownership as a way to have the former in the resource sector while avoiding the danger of the large corporation. But Nelson also notes that this tradition has been supplanted since New Deal days for the most part by appeal to what Theodore Lowi has called "interest-group liberalism" in which political pressures rather than the apolitical criteria of scientific management control land use. Moreover, he is not sanguine that scientific management, if definable, is as likely to be achieved under public management as it is in large corporations where control is insulated more effectively from politics.

Whatever the validity of the original rationale for retention, it appears to have been greatly attenuated by subsequent events. Private operators have acquired management expertise, and they now are motivated by market forces to plan for the long term to protect their investment. Those who favor retention of the federal lands need to state their case in somewhat different terms for present circumstances.

The present composition of the federal lands includes much grazing land that historically was unsuited for crops. The other principal component is forestland reserved in the heyday of the conservation movement or acquired later under the Weeks Act. Perry Hagenstein provides a concise review of the present use and status of federal lands in chapter 4. Special conditions prevail in Alaska, where land issues recently were legislated after lengthy consideration. Other federal lands are held as national parks, fish and wildlife preserves, military reservations, or for other purposes, and some backcountry land is dedicated for trails, wild and scenic rivers, and wilderness. Thus, only about 285 million acres remain in the multiple-use categories where commodity and amenity uses compete. For the most part, federal lands are not of highest quality for commodity production, and the best scenic and recreational resources are not found on multiple-use lands.

Hagenstein finds the federal lands limited in the contribution that they can make to society. They supply about one-quarter of the nation's softwood timber harvest and only 2 to 3 percent of forage for livestock. Their importance for minerals is a more complex issue, but federal onshore lands supply about 5 percent of U.S. oil and gas production, 10 percent of coal output, and a store of hardrock minerals that pass into private hands when the deposit is exploited. And they also supply extensive outdoor recreational opportunities. The uses are largely private, and conditions of access vary from one commodity or use to another. In the case of timber and some minerals, as well as recreation, the supply made available by federal managers can significantly affect market prices.

While the government has the power to use its lands as it sees fit, restrictive arrangements of one kind or another strongly condition this power in practice. Hagenstein details some of these arrangements. They favor miners and ranchers, allow virtually free recreational use, concede control over wildlife and water to the states, and in general, benefit specific interests and local populations. These arrangements are embedded in law and custom, and change is an arduous process. The requirements of the National Environmental Policy Act (NEPA) and agency planning procedures can serve to paralyze action that might change present land use sharply. Reviewing changes emerging from the work of the Public Land Law Review Commission (PLLRC), Hagenstein offers the gloomy prognosis that history and special interests will frustrate efforts to move either in the direction of major internal reform or toward wholesale disposal.

In commenting on Hagenstein, Michael Harvey, in chapter 5, places somewhat more stress on the value of federal lands, including values that are not recognized in the price of real estate. He sees growing competition for the use of increasingly valuable lands and generally believes that the balance between commodity and environmental values on the land can best be sought through active administration of the public land laws which, he argues, allow for full representation of all views. Harvey recognizes that there have been administrative deficiencies—in particular, he would hold public managers to stricter accounting of capital items—but his basic faith is that it is possible to improve the workings of the present system so that it will serve us well if only we all behave in a restrained and reasonable manner. There is in this comment an almost classic merger of the views of progressive conservationism and interest-group liberalism that appears to predominate among practical conservationists.

Part III attempts to present the kind of intellectual arguments or rationale that can be offered either for retention or disposal of the federal lands. The chapters in this section offer a sharp contrast in views. In chapter 6 Joseph Sax suggests that the matter of ownership should not be overplayed. Government could, if it wished, manage the land much as a private owner would do and perhaps would do so with equal efficiency (a point disputed by his opponents). Moreover, in its extensive power to regulate, government could achieve much the same results as through ownership, though he thinks there may be tactical advantages to ownership in this regard.

To Sax, the debate over ownership is mostly of symbolic significance. Those who would sell basically are unreconciled to the newly emerged environmental values reflected in the Federal Land Policy and Management Act (FLPMA) and other recent public land legislation, while their opponents distrust the trend of the Reagan administration, and former Interior Secretary Watt in particular, to the extent that even seemingly very minor and justifiable sales are seen as part of a larger scheme favoring a still unreconstructed private sector heedless of the environment.

Even more fundamentally, Sax sees the debate as part of the larger controversy over collective versus individual values. This problem transcends the question of ownership, since again collective values can be pursued through regulatory devices as well. To market adherents, individual preferences are sacrosanct and they are best expressed through markets; a well functioning market yields the correct outcome. But Sax argues that people also hold preferences or values in their capacity as members of collectives and that those values cannot be expressed except collectively. Both government ownership and regulation may express a form of collective preference.

Much of Sax's argument is devoted to establishing that collective preference, while different, is equally as legitimate a form of preference as the usual individual preferences expressed via markets. Collective preferences might more aptly be called community values that command loyalty at the expense of individual autonomy. Is individual autonomy the "premier virtue," Sax asks, or is it not the case that people also have the urge to subordinate some part of themselves to external moral authority and to draw strength from belonging to collectives, whether club, church, or state? Most people live somewhere between the absolutes of autonomy and slavery and see values expressed collectively as a normal part of their lives.

Collectives have pragmatic as well as psychological significance in this formulation. They may be an administratively efficient way of accomplishing shared objectives. They may be an efficient way of getting information. They may help an individual to shape his own reactions. Thus, even when his participation is voluntary, membership does not merely reflect, but also shapes, the individual's view.

The state, of course, is an involuntary collective and it has coercive power. Sax observes that some formally voluntary organizations, such as churches, involve entanglements and commitments whose rupture would entail enormous psychic cost, so the distinction between voluntary and involuntary collectives is not as absolute as it appears. Citizens have accorded the state power over a long period of history and have not insisted on such stringent limits to the state as to maximize individual autonomy. Rather, the approach has been to define personal rights not subject to state coercion instead of designing a minimal state in all spheres.

There are collective values, such as protection for the rights of a future generation, or existence values that may be neglected by self-seeking individuals. These values cannot be measured by the standard of individual preferences. The state can address these values and, through public action, can assure that free riders do not defeat the common interest in providing for them; that is, the state can provide them where the market would not. The objectives of government action often are more complex than those of a private action, and there may be no convincing criteria by which the wisdom of the action can be judged. However, they are subject to evolution over time, and in a democratic society the political process favors some review and restraint on inefficiency, corruption, and special interests.

Although it is control over land rather than ownership which determines whether land will yield the public values of interest, Sax believes that retention of federal lands is viewed as a means of ensuring the success of public values when they diverge from private interests. The past two decades have seen a substantial redefinition of the line between private rights and public values in favor of the latter, but the new line is not yet fully secure. Hence the sale of the lands would be viewed as a step away from that newly established balance, even though it need not be such; the symbolism of the move becomes important.

Richard Stroup, in chapter 7, allows that collective action is legitimate but questions whether it will prove to be the best alternative in most cases. Neither markets nor governments are perfect, but bureaucratic management often defeats both economic rationality and the collective values it is intended to protect. Private parties and voluntary organizations show environmental sensitivity and greater agility in seeking it. Collective choices, in his view, are more likely to be ill-informed and governed by special interests than are private decisions. He argues that environmental interest is not the same as the collective interest and that, with some institutional adjustments, the environmental interest can be reconciled more readily with economic values through the use of markets.

This point of view is developed more fully by B. Delworth Gardner, who makes an argument for aggressive divestiture of federal lands. He rests his case more on efficiency and equity grounds than on ideological argument, though some might contend that his very acceptance of the economic efficiency paradigm is an ideological choice which foreshadows all that follows.

Gardner, in chapter 8, criticizes public management as failing in all dimensions of efficiency—it produces a less than optimal output from the land, the quantity and productivity of public investment is too low, and the distributional results are questionable. These outcomes are judged by the efficiency norms of private markets. He admits, however, that the criteria for equity are more ambiguous than those for production and investment.

The reason that public managers do not match the efficiency results of private firms is that they must respond to political decisions on how to allocate federal land resources rather than to prices. Moreover, they do not have the information that prices convey, and, without prices and market discipline, they lack the incentive to acquire that information. Political bargaining is not the equivalent of price in achieving efficient allocation. Instead it invites the contestants to seek power and influence so as to claim the rents available from federal lands. In the contest, many of those rents simply are dissipated in a social sense by the competing efforts to capture them. The bureaucracy also becomes a player in this unproductive game, seeking to defend its own position rather than to pursue efficiency in a single-minded way. Administrative principles such as multiple use and sustained yield are not efficiency rules: instead they become means for validating the political allocation that actually occurs. On the investment side in particular there is little chance to achieve optimum results, since the agencies are dependent on appropriations not governed by efficiency and on fees that are too low, that go mostly to the states, or are allocated by formula rather than in relation to economic yield. About the only positive aspect of public management in Gardner's view is that it permits the consideration of third-party interests, which the market does not.

Gardner offers a number of examples of inefficient public management. They include the overgrazing of Indian land, unjustifiable chaining of grazing land, adherence to allowable-cut rules, and rules restricting new stockmen from competing for public forage. While these cases are documented, they may not prove the case to everyone's satisfaction, for it would not be difficult to compile a similar list of private mistakes as well.

His positive case for disposal rests very largely on a portrayal of the outcome to be expected from private markets. Private managers have the incentive to acquire information, reduce unit costs, and optimize product diversity. Moreover, they will invest to increase output and, being optimists, will want to conserve the future value of their assets. It should be noted that this represents an idealized picture of the way that private markets operate rather than an actual detailed description of their performance. It is conceded that public goods are not accommodated well by private markets. However, some public good will be produced on private lands, including the existence values and watershed benefits that Sax cherishes, and even more could be assured by imposing restrictions at the time that title to the land is transferred. Also, many presently unpriced goods could be priced and, therefore, might be produced. Gardner finds it unlikely that public managers would do as well because they are subject to capture by the interests that they seek to control.

Although Gardner asserts an equity case for divestiture of federal lands, he does not develop it at much length. The lack of well-accepted equity criteria within the economic efficiency paradigm that he accepts does not permit him to go very far in that direction. However, he is able to cast doubt on the equity outcome of federal management by citing cases in which it seems perverse. Since a major reason for government action always has been the promotion of equitable outcomes, his examples seriously cloud that basis for holding land in public ownership.

Gordon Bjork, in chapter 9, argues that factor price shifts originating in economics and technology occasion the redefinition of property rights. In present circumstances such changes favor a shift to private property rights, as he sees it. Most of his discussion concerns forestlands suited to the production of commercial timber. Developments that improve yield, reduce growing and harvesting costs, and increase the relative price of timber all favor timber cropping and harvesting of old-growth timber in line with the intensive management practices used on private

land. Of course federal managers in theory could adopt similar management practices, but they are inhibited from doing so by bureaucratic
and political considerations. In Bjork's view, the shift to private rights
would increase the economic rent that the land yields. While public
ownership might offer better control of externalities and more output
of public goods, he is not convinced that these are significant advantages;
private management will provide many of the same values in any case,
and the magnitude of those lost should not be overestimated.

Bjork estimates that the opportunity cost of increased commodity
production on federal land is on the order of $100 per capita annually.
This is the commercial output that would be forgone by continuing
present amenity and environmental uses.

In Part III, more practical possibilities for tenure changes are considered, along with some of their implications. In chapter 10 Clawson
ranges widely across the subject of federal lands, noting the changes in
their status over time and recent changes in their value, in the dispersion
of management expertise, and in political concerns that warrant an
examination of possible new tenure arrangements. He considers several
possibilities for change that might be used in some combination. However used, they are likely to be applied selectively and over an extended
period.

The least drastic approach would be to retain the lands but seek
improved management. Much of the recent land-management legislation
aims in this direction. A first step would be to set up a capital accounting
system to allow capital values to be reflected in decisions. Another would
be greater use of fees and charges, especially for recreation, and allow-
ance for more year-to-year flexibility in management. All of these steps
aim at increasing economic efficiency in federal land use. They leave
open the question of whether the same purposes might better be sought
privately.

Clawson is very skeptical about transferring federal land to the states—
the core proposal of the Sagebrush Rebels. He traces a long record of
state ineptitude in the management of state lands and notes that state
ownership preserves the public management so often condemned when
exercised at the national level.

Historically, outright disposal was the common expectation for federal
lands. By 1920, virtually all of the land capable of producing crops had
passed into private hands, but the law was unsuited to disposal of forest
or grazing lands in tracts of economically viable size. Clawson believes
that much federal land now would be bought by ranchers, the forest
industries, miners, and conservation groups if it were made available in
appropriate units. Since public surface uses have not generated revenues

sufficient to cover costs, nor are they likely to in his view, from a financial standpoint the government would well be rid of them. But the process would be selective, and much unproductive land would not find buyers; the government long will remain a substantial landlord. The idea of public corporations has been advanced as a way of managing federal lands. They also could serve as transitional devices for conversion to private holdings. Great flexibility is possible under this concept, but Clawson remains skeptical that such corporations would retain their élan over an extended period of time. In their aim of injecting more economic rationality into federal land management, public corporations do not appear superior to private ownership.

Finally, Clawson devotes major attention to a scheme for long-term leasing of federal lands. Leasing is widely used in the case of private lands, and that experience guides the proposal for federal leasing. By specifying the terms of use, providing for early renewal, and compensating for unharvested timber and the like, the government grants a form of tenure that invites investment and intensive management. Clawson would open lands of all types—forests, grazing lands, and recreation lands—to prospective lessees on a competitive basis. In practice, only the most attractive lands of each type would be sought by lessees. The size of unit would be of economic scale and of sufficient duration (one hundred years for timberlands) to justify good management practices. Those lands not capable of an economical level of commodity production could be leased by conservation groups or private parties for nominal charges subject to certain restrictions on use. State and local governments also could compete for leases.

A novel feature of Clawson's leasing proposal is the provision of what he calls a "pullback." This would permit rival claimants who were willing to meet the same terms to take a portion of any proposed lease. It would impart elements of competition and negotiation into the leasing procedure. A proposed lessee would have an incentive to come to terms with his adversary rather than to dump the problem on the courts or the agencies. Private bargaining and contractual arrangements analogous to those in private markets would replace much of the current dependence on the political and bureaucratic process.

In chapter 11, John Leshy, approaching his subject from the standpoint of law and politics rather than economics, notes the great inertia of the present federal lands system and details some of the specific issues that would be raised if any major change of status were contemplated. This dose of reality bears close reading, for its burden is that the present balance between interests is politically rather stable, whatever its economic rationality, and that threatened changes are likely to be defeated by coalitions of those who see themselves as potential losers.

Leshy observes that the government has legal power to manage or dispose of its lands pretty much as it sees fit. Indeed, already it has experimented with the widest range of devices for transferring title with or without restrictions and transferring partial title, as well as surrendering administrative control over some resources—especially wildlife and water—to the states. But he finds that restrictive transfers of title are very much bound up with the political process and with the political will to enforce the restrictions. That, in turn, is reflective of the political status quo, almost irrespective of the wording of the law. Thus, graziers held effective veto over federally reserved coal beneath their lands even before the law was changed to recognize that fact, and provisions allowing mineral entry on wilderness lands until 1984 have been rendered ineffective by the power of the environmental movement. Any movement toward reduction in federal control would create its own inertia and impede attempts at continued federal surveillance.

It is with regard to such special aspects as mineral or water rights that a change of tenure would arouse the most concern. Would hardrock miners acquiesce to changes in status that almost certainly would imperil their access to federal mineral deposits and put renewed strength behind proposals for a comprehensive mineral leasing law? Leshy is skeptical. In the case of water, the federal government has largely abdicated its rights and control over water originating on federal land—about two-thirds of the total in the eleven western states. If changes in the status and use of federal lands increase demand for water or decrease the supply and quality, they will place a severe burden on the delicate structure of western water law. Both present water users and state authorities would be skittish over such changes.

Leshy notes that western states' "rights" to control water and wildlife lack constitutional basis and are protected only by the political process. The western states have entrenched power over federal land-management matters, and the attitudes of westerners are vital to any proposals for change. The western states already get most of the mineral revenues generated on federal lands and are overcompensated by payments in lieu of taxes for those services they provide to federal lands. He suggests that much of the recent agitation may be aimed at gaining still more favorable treatment for western interests rather than at attempting to rupture the federal link.

He concludes by suggesting that most participants view federal land management as a zero-sum game where one profits only at the expense of another rather than as a positive-sum game capable of generating larger benefits. They remain to be persuaded otherwise, and many can exercise effective vetoes over change via the political process. He sees stalemate.

Discussants of these papers were asked to comment not so much on the technical aspects of the main papers as to present different perspectives on the subject. In a paper not reproduced here, F. L. C. Reed provided a Canadian perspective on public lands that was in marked contrast to much of the discussion about the United States. About half the land in Canada is owned by the provinces and another 40 percent by the federal government (mostly in the north), leaving only 10 percent in private hands. The provinces own the minerals and most of the forests in their territory. Despite this great concentration in public ownership, there is little pressure to dispose of the land. Rather the emphasis is on improved management of public lands, in part through long-term management agreements with private firms. The latter are expected to regenerate areas cut, and renewal of their contracts is based on performance. However, benefits are not judged by financial criteria to the same extent as in the United States; rather, more weight is given to economic, social, and environmental matters. Reed stated that an attitude of stewardship toward the land substitutes to some degree for the more demanding economic criteria prevalent in the United States. Although Reed did not discuss this, much of the Canadian crown lands are of low value, making for less disposal pressure and favoring stewardship.

Robert Nelson's discussion provides a fitting final chapter for this volume. One of his thrusts is to provide a context of intellectual history for the argument about federal lands. Disposal of land throughout the nineteenth century fitted the prevailing ideology of classical liberalism with its preference for the private sector and limited role for the state. Retention of land in public ownership reflected the progressive ideology of the turn of the century, with its belief in scientific management and longer-range view. Subsequently, efforts to balance competing interest groups in multiple-use management reflected the active application of "interest-group liberalism." Despite the important differences between progressivism and interest-group liberalism, the two ideologies often are invoked together in defense of government policy. Current interest in the disposal of land betrays some loss of confidence in the capacity of government to solve problems. He finds that there has been little serious recent examination of why some lands should be public and others private.

Nelson's main criticism of interest-group liberalism is that its stress on compromise usually is inconsistent with efficiency or effective planning. It also encourages the zero-sum conflicts that Leshy laments. It magnifies the strength of concentrated interests at the expense of the more diffused.

In reviewing Clawson's alternative prescriptions for the federal lands, Nelson provides a valuable critique of each. Those who would improve government management approach it from two different and inconsistent points of view; one favors greater insulation from politics through the use of economic criteria, and the other moves toward more political involvement and public participation. Neither offers much promise, according to Nelson.

Those favoring sale of federal lands see scientific management by government as an illusion and consider market results, while not perfect, a major improvement. Nelson is sympathetic to this view and sees large corporations as the modern embodiment of the progressive ideal of scientific management, especially for commodity-producing lands. However, he recognizes the importance of nonmarket values on federal lands, especially recreational uses and existence values, and sees a role for public (not necessarily federal) ownership in these cases.

Nelson is more receptive to an enlarged role for state ownership of land than most. States already possess many of the attributes of ownership over federal land in terms of revenues received and through the application of state law—perhaps they should be responsible for the costs as well. The past record of poor performance need not be prologue in the changed circumstances that would be created by transfer. Much ordinary federal land is used mostly for recreation and mainly by local residents.

After showing little enthusiasm for public corporations, Nelson turns to long-term leasing. While leasing appears not to have the finality of sale, the differences "are largely symbolic" in the case of a one hundred-year lease. Nonetheless, it might be politically more palatable than other changes and could provide for the kind of private incentives that are sought. His specific reservations about Clawson's proposal are better read than summarized.

Returning to his theme about the role of ideas and ideology, Nelson suggests that the federal land system may not be as immutable as Leshy implies. He contends, "If an intellectual consensus were to develop for major changes in public land tenure, I have the feeling that the political process would follow." His favored solution would classify lands according to principal use and allow for different tenures for each. Thus, recreational lands would be candidates for public ownership, while commodity lands would be better suited for private management. This involves a compromise among broad ideologies rather than narrow interest groups.

At the symposium, individuals selected from groups with an interest in the federal lands were invited to respond to the papers; a general discussion followed. Those responses are not reproduced here, but from text and tape their general flavor can be reconstructed.

Unsurprisingly, neither of the spokesmen from federal land-management agencies—Ray Housley of the Forest Service and Frank Edwards of BLM—was much captured by suggestions for major changes of tenure. Both were careful to stress the limited nature of the Reagan administration's program for land sales, with Edwards noting its long history in the bureau's planning process and the basis for it in law. Housley conceded that the government might not own all of the right acres, but the Forest Service program has initially identified very few parcels for sale; much of the rest is in a category for further consideration, with the frank expectation that most of that will be retained. While Housley was intrigued by Clawson's leasing proposal, he clearly took comfort from the views of Harvey and Leshy that major changes are not likely and was prepared to soldier on at improving public land management.

Rowena Rogers, president of the Colorado Board of Land Commissioners, was drafted to speak from the perspective of state and local government. She noted the dominant role that federal land plays in many rural areas of the West and the sharply curtailed tax base that many local jurisdictions have for servicing the activities on federal lands, ending up with a passionate defense of federal payments in lieu of taxes as an essential support for service to federal lands. Rogers was undogmatic on the question of tenure but pointed out that the states have the capability to manage lands and that some federal lands long due to the states remain undelivered. A recurrent theme of her remarks was the local sense of powerlessness—that decisions vitally affecting western communities are made elsewhere without full consideration of local interests. Therein lie the roots of the Sagebrush Rebellion, in her view.

Rancher Laird Noh seconded much of Rogers's view about the Sagebrush Rebellion, endorsed the limited version of the asset-management program, and then went on to confront the broader issue of privatization. Noting that the livestock industry is philosophically in tune with that idea, he nonetheless sounded a note of caution. He remarked that the industry has about $1 billion invested in existing grazing rights on public lands and, if those assets are threatened, then the industry and many local communities would face ruin. He would like to explore possible ways of increasing the private stake in the present public lands so as to improve their management but was even more concerned about maintaining some stability and consistency in policy. His was a cry on behalf of the working rancher who faces all of the hazards of his business and shudders at the thought of drastic policy changes that might wipe him out.

George Craig spoke from a background in the forest industries. Craig sees a role for both public and private management, but feels that the timber industry is discriminated against in the management of federal land because it is the only user expected to pay a fair market value for its use of the resource. This situation, being associated with government ownership, institutionalizes inefficient practices. Craig cited numbers to support his argument that federal forestlands are inefficiently managed and focused on the excess inventory of overage timber stands as further evidence. He was not sanguine about the prospects for a leasing scheme, arguing that preservationists and bureaucrats attain their purposes under present arrangements without being forced to compete with others, as leasing would require. Moreover, the forest industry itself is split over the advantages of changes in tenure. Rather surprisingly, he included a plea to allow the RPA process to work, perhaps because he is pessimistic about tenure change and sees this as a path to improved efficiency.

The final remarks came from former U.S. Senator Gaylord Nelson, now chairman of the Wilderness Society, who presented a spirited rebuttal to the case for privatization. In response to those who argue that the private interest will provide protection for the social interest in the resource, Nelson pointed to extensive soil erosion on private land and to the record of degradation in private eastern forests that imposed costs on society through erosion and floods. He found privatizers insufficiently concerned about the nonmarket values of public land and was unconvinced that private devices could substitute for public management of these values. He noted that others share that estimate, for wealthy individuals sometimes give land to the public for the purpose of preservation. Recreational values in particular would suffer if the land went into private hands. Moreover, he saw little need for a change of tenure because private interests already produce the commodities from public lands, and public ownership is needed to ensure the other outputs. Finally, he rejected the notion that the environmental view represents a narrow, elitist, special interest by citing the widespread support that the movement still commands.

Fuller discussion of most of these points can be found in the chapters that follow. In sponsoring this discussion, Resources for the Future has not intended to take any position on the merits of the argument. However, we do feel that it is entirely appropriate to give reasoned consideration to basic issues concerning the public lands. In that spirit, the reader is invited to examine the fact and argument that follows and to make his own judgment.

Notes

1. Robert A. Nelson and Gabriel Joseph, "An Analysis of Revenues and Costs of Public Land Management by the Interior Department in 13 Western States—Update to 1981" (Washington, D.C., Office of Policy Analysis, U.S. Department of the Interior, 1982).

2. Richard Nelson, "Assessing Private Enterprise: An Exegesis of Tangled Doctrine," *The Bell Journal of Economics* (Spring 1981).

3. John V. Krutilla and John A. Haigh, "An Integrated Approach to National Forest Management," *Environmental Law* (Winter 1978).

4. John V. Krutilla, Anthony Fisher, William F. Hyde, and V. Kerry Smith, "Public Versus Private Ownership: The Federal Lands Case," *Journal of Policy Analysis and Management* (Summer 1983).

II
Federal Lands
Why We Kept Them
How We Use Them

2

The Federal Lands
Why We Retained Them

PAUL W. GATES

The foundation of the Public Domain of the United States dates from the Treaty of Peace with England in 1783, when that country recognized the independence of the United States and ceded its claim to land south of the Great Lakes, and from the cession by the states of their western land claims.[1] Congress asked the states to convey to it these claims because they overlapped and were a constant source of dispute and because the lands would be a source of revenue and power for the new federal government. In making its cession, consisting of the five old northwestern states north of the Ohio and east of the Mississippi, Virginia declared that the land "shall be considered as a common fund for the use and benefit" of all the states and "shall be faithfully and *bona fide* disposed of for that purpose and for no other use or purpose whatsoever."[2] Congress, in turn, pledged that all lands ceded by the states would be disposed of for the common benefit of the United States, thus making it clear that the public lands belonged to all the people. The Northwest Ordinance of 1787 and the new federal constitution of 1789 provided that new states were to be admitted into the Union "on an equal footing with the original states in all respects whatever." But unlike the original states, which had retained the public lands within their boundaries, the new states had to accept federal supremacy over the public lands and to agree not to tax them and never "to interfere with the primary disposal of the soil by the United States in Congress assembled, nor with any regulations Congress may find necessary for securing the title to such soil to *bona fide* purchasers."[3]

Few questions of more vital importance faced the new nation than deciding what was to be done with the 233 million acres of public land it acquired between 1780 and 1802. Should the United States make extensive land grants as England had to William Penn, General Oglethorpe, and the Hudson's Bay Company? Or should the area be divided into states, with the states receiving those public lands within their borders as England had done for Massachusetts and Connecticut? Should the land be retained by the United States while the country was encouraging the development of self-governing states under the suzerainty of the United States? Should the public lands be reserved as a source of revenue to aid in paying off Revolutionary War debts and in keeping taxes down, or should they be opened to settlement and made easily available at land offices where settlers and investors could purchase them? Or, should they be free, with the owner paying but a small fee to secure title after the land had been settled upon and improved?

Two prominent Americans were to have a profound influence on government policies for the new territories and for the sale and development of the public lands.[4] Thomas Jefferson, son of a well-to-do Virginia planter, was a firm believer in democracy and in the democratic distribution of the public lands to landless people, both those living here and those drawn from abroad by the promise of democratic institutions. Democratic distribution of the public lands to the landless so they could become independent farmers would, he believed, assure the preservation of the American republic and avoid the excesses that might arise from a turbulent population that had no stake in the land. Alexander Hamilton had married into a wealthy family that owned a great estate cultivated by tenants. He was closely associated with rich merchants and land company officials who were deeply interested in speculating in land. From these associations Hamilton came to believe that land investors and stock companies were essential in drawing people to the West by their investments in roads, mills, and factories and that the government should sell its lands in large blocks to such men, letting them act as middlemen between the government and the settler. Both Jefferson and Hamilton, however, believed it wise to use the early revenue from land sales to retire the Revolutionary War debt.

The influence of their opposing ideas created an interplay of forces between the speculator-promoter and the endless stream of people seeking to acquire small quantities of farmland. Never was the United States to choose which element it considered the more useful.

Neither Jefferson nor Hamilton was present when the Confederation Congress established the basic structure of the American land system and the system of territorial government. The Land Ordinance of 1785 was borrowed mostly from the New England land system with its rectangular surveys into townships of 36 square miles and sections of 640 acres, with section 16 reserved for schools. Except for inaccessible areas and those private land claims dating prior to American control, the public lands have everywhere been laid out on the rectangular system, regardless of whether it was well adapted to the area. The Northwest Ordinance of 1787 prescribed the form of government for the territory beyond the Ohio River, lessening congressional control in three successive stages as their population grew. In the third stage, admission into the Union was granted on the basis of equality with the original states. All public land states, except California (the original thirteen states plus Vermont, Maine, Kentucky, Tennessee, West Virginia, Texas, and Hawaii were not public land states), came up through this territorial system, and all from Ohio in 1803 to Alaska in 1959 had to agree "never to interfere with the primary disposal of the soil by the United States. . . ."

One of the first steps taken by Congress during the Washington administration was to include in the Public Debt Act of August 4, 1790, a section declaring that the proceeds from the sale of public lands "are hereby appropriated toward sinking and discharging the debts . . . ," doubtless the result of pressure from Hamilton.[5] Both Hamilton and Jefferson were fearful of the influence that holders of the public debt might have on the government and looked forward to its final retirement, which came in the 1830s. Hamilton's views prevailed in Congress; the price of land was set at $2.00 an acre, and the smallest tract to be sold was 640 acres. Before the Ohio country was opened to settlement, squatters had intruded. They were ruthlessly ejected from lands they had taken illegally and their improvements were destroyed. Congressional will could prevail in law, but squatters from Virginia and Kentucky could return to their holdings, and did so. Later they induced Congress to grant them preemption rights.[6]

The government had taken a revenue stance concerning the public lands but the revenue from land sales was negligible in the early years. Not until 1818 did it amount to more than 10 percent of government receipts. When public land sales began to produce large sums—constituting 7 percent of federal income in 1811, 13 percent in 1819, and 48 percent in 1836—there was no general movement to reduce the price of land, except for land that had been passed over as undesirable.[7] Instead, Henry Clay won great popularity for himself by favoring the distribution of net proceeds from the sale of public lands to the states on the basis of their representation in Congress instead of lowering either the tariff or the price of land to cut the government's surplus revenue.

Under the terms of Clay's Distribution–Preemption Act, the non-public land states received two-thirds of the amount distributed before distribution was suspended, the other third going to the eight public land states.[8] But each of the public land states also received 500,000 acres for internal improvements. Most important, this measure gave the West its first objective, permanent, prospective preemption for settlers. Thus from the admission of Ohio into the Union in 1803, Congress granted land to the public land states and liberal concessions to settlers which were not shared by the older states, thereby doing violence to the principle that all the states should be treated alike.

Lacking the financial means to induce sufficient men to enlist to fight in the Revolutionary War, and later in the War of 1812 and the Mexican War, Congress rewarded soldiers with land in military districts in Ohio, Michigan, Illinois, Missouri, and Arkansas. These grants amounted to 78 million acres. Few of the soldiers of these three wars and of the Indian wars were inclined to go west to take up their claims as Congress had hoped. Consequently, the warrants (mostly from 40 to 160 acres) were dumped on the market and sold for 80 cents to $1.00 an acre. In other words, the military warrants became virtually land office money, and because of their volume they fell in price, thus lowering the cost of land and the revenue the government might have obtained.[9] Here is an early instance of the willingness of the federal government to part with its land for national purposes.

Aside from Clay's distribution bill, only in three later instances did the eastern states gain direct benefits from the public lands—by the two Morrill Acts of 1862 and 1890 and the Hatch Act of 1887 which aided the states in creating experiment stations. Clearly Congress had forgotten the pledge to the older states, but these states were never to forget their surrender of the western lands and their right to share in the benefits from them. Although, after numerous refusals, they ceased to ask for direct grants for canals and railroads, they clung to the notion that the public lands belonged to the people of all the states.

Meanwhile reform-minded people, aware that donations of land would be much easier to get from Congress than money, brought forward many schemes for sharing the lands with the states for colleges, seminaries, and institutions for the care of the blind, the insane, and other unfortunates as well as for a host of miscellaneous purposes. Congress would not agree to raise money by taxes for any such purposes, but if land were given for common schools or for roads, it could be argued that the establishment of these facilities would add to the attraction of the remaining public lands and bring about better prices for them.

With this rationalization, Congress set happily about providing subsidies, that is, sharing the public lands with the states created out of the public domain. The largest of such grants was the allotment of the sixteenth section in every township for schools. The school grant had been adumbrated in the Land Ordinance of 1785 and definitely provided for Ohio in 1803. Thereafter, every state carved out of the public lands received one-thirty-sixth of all public land in the state for schools. In 1846 Congress doubled the grant, and in 1894 it granted four sections— or one-ninth of the land—to Utah. Later this also was done for Oklahoma, Arizona, and New Mexico. One can imagine the angry denunciation that proposals to use federal money for the support of common schools would have aroused at any time in the nineteenth century. Other grants were given for religion, for the New Madrid Earthquake sufferers, for the Moravian Indians who had been brutally treated by squatters, for Polish exiles, and for an asylum in Kentucky. And a township in Florida went to General Lafayette for his aid in the American Revolution. Thomas Donaldson, the compiler of public land data, estimated that a thousand such measures had been adopted by Congress and that tens of thousands of such schemes asking for donations of lands were only resisted by "a jealous care on the part of Congress. . . ."[10]

In the opinion of most westerners, land grants for education were acceptable, but grants to improve transportation facilities for shipping their grain, livestock, and cotton to market would be even better. In the 1820s Congress experimented with land grants for roads following the alternate-section pattern so familiar to most westerners today. Then followed grants for canals and railroads, all bringing about the transportation revolution that George Rogers Taylor has described so effectively.[11] The alternate-section pattern was a modernization of the argument previously used for education grants, that one-half of the land within the granted area would bring as much after the road, canal, or railroad was constructed as the entire area would bring without the grant.

Between 1785 and 1870, Congress had gone far in using the public lands for causes that middle-of-the road constitutionalists thought proper, while avoiding any direct money grants, though the distinction between them was becoming less clear. At a time when it was difficult to gain appropriations for national social and developmental purposes, the ease with which support could be obtained for grants of land for these purposes was one of the nationalizing forces in American economic development.

Though the original eastern states had hoped that the income from the public lands would be a major source of revenue for the government and free them from heavy taxation, these states were liberal in supporting grants of land to the western states. It was expected that the lands would be sold and that funds received would finance the purposes for which the grants were made. For example, income from the school sections was for the support of education, and money from the sale of the agricultural college lands was to be invested as endowment for the colleges. The 64 million acres of swampland given the states would, it was hoped, provide sufficient funds to enable them to undertake important drainage projects. However, the record of the early states in disposing of their grants was so characterized by mismanagement that Congress later turned to conditional grants requiring that the land should not be sold for less than prescribed sums.[12] In 1876 Colorado was required not to sell its education grant for less than $2.50 an acre, and in 1889 and 1890 land in six new states (North and South Dakota, Washington, Montana, Idaho, and Wyoming) was not to be sold for less than $10 an acre. This limitation was commended by Dakotans, fearful that the income from the school sections would be lost quickly if local administrators rather than state officials were to have jurisdiction over the sale of the land. The $10 minimum tended to keep a larger proportion of the land in state control than would otherwise have been the case.[13]

If we exclude certain southern politicians of the antebellum years, we can say that representatives of the older states generally favored a liberal land policy, although they would have liked to share in the benefits the land grants were bringing to the West—increased population, investments, and rapid economic growth. In only one instance had they shared in these land grants and that was by the Agricultural College Act of 1862. This act was preceded by an extraordinary bill providing for 10 million acres to be divided among the states for improving the care of the insane. It swiftly passed through Congress, the result of lobbying by one of the ablest reformers of the nineteenth century, Dorothea Dix. She presented the issue first to the states, where by effective lobbying she secured instructions from state authorities to their congressional delegates urging them to vote for her bill. It worked; the bill passed both houses by large majorities but President Pierce, who had apparently escaped the intimidating effect Miss Dix had had upon others, chose to veto it. It proved impossible to pass the bill over his veto.[14] The incident's significance lies in the fact that a similar method of granting aid to all the states for the establishment of agricultural colleges was taken over by Justin Smith Morrill of Vermont. In 1862 Representative Morrill won the enactment of what came to be known as the Morrill Agricultural

College Act and his name was to be perpetuated on a number of state college campuses where buildings were named after him.[15]

Since it would have been unwise to allow eastern states to own land in western territories or states, the Agricultural College Act required that the grants be given to those states in the form of scrip to be sold by them and entered for land only by the purchasers of the scrip. But this use of scrip played into the hands of speculators, just as the large quantity of military bounty warrants had enabled capitalists to enter large amounts of land in the West at low prices. Morrill's Act granted 30,000 acres to every state for each representative and senator it had in Congress, thereby assuring that New York, Pennsylvania, and Ohio would get large grants (990,000, 780,000, and 630,000 acres, respectively), whereas new states with small populations would get only 90,000 acres.[16] Three of the eastern states evaded the provision requiring the scrip to be sold by conveying it to the colleges they created and allowing them to enter the scrip. With New York's land scrip, Cornell University acquired a half-million acres of pine land in Wisconsin from which it was to make one of the most successful speculations of the nineteenth century. The Morrill Act was the only act granting land in the West to eastern states.

Having found a way by which the East could share the wealth provided by the public lands, Morrill again tried to get help for the agricultural colleges on the ground that states with a small representation in Congress—and that included, in addition to Vermont, most of the western states—had derived from their grants little more than was sufficient to endow a single professorship. Since the West was adamant that no more land should be given to the eastern states because the scrip they had received had been used largely to enter first-rate agricultural land in California, Kansas, and Nebraska, thereby reducing the amount of land that would be available for homesteads, Morrill advocated that a grant of money be given to each state out of the income from land sales, which would be used to supplement their meager funds for their land grant A&M institutions.

The scheme won sufficient support and, in 1890, the second Morrill Act passed, providing a grant of $15,000 annually to each state plus an additional $1,000 every successive year until the annual grant amounted to $25,000. Three years before, Congress had authorized an appropriation of $15,000 annually from public land sales to aid in establishing agricultural experiment stations in each state. Western interests soon realized the danger in allowing the nonpublic land states to share in the income from public land sales, and in 1902 all remaining income from such sales in the semiarid states was to go for irrigation development.

Congress had then to declare that appropriations for the agricultural colleges and experiment stations should come out of general revenues.[17] Morrill had won a public land grant and income from public land sales for his colleges, but his method was never tried again. The Newlands Reclamation Act had forestalled any such efforts. The West had made certain that every last cent of net income from public land sales was to be distributed to the states in which the sales had been made and used for reclamation in the semiarid states.[18]

The irrigation projects of the Reclamation Service were as great a departure then as the TVA was to become at a much later date. They modified the pattern of agriculture and transformed or created new communities. The government was charged with impounding the waters of major streams in large reservoirs, conducting the water to irrigable lands, and exacting payment for the water in the form of rents spread over many years. It would soon have hundreds and thousands of debtors, who over the years obtained increasingly liberal treatment in repayment clauses that seemed to approach outright donations.[19] What the federal government had done for the industrial East by the protective tariff, it was now doing for the West through the Reclamation Service, making new western communities the fastest growing in the country. Westerners caught up in the nostrums of the Sagebrush Rebellion should reflect on how much the development of their communities owes to the low-cost water and power that the Reclamation Service provided through federal control of public lands.

In agreement with the Jeffersonian notion that landownership in a republic should be widely distributed, Congress, in 1862, took what appeared to be the final step in implementing this ideal. It promised any citizen, or prospective citizen, title to a free homestead of 160 acres of surveyed land after five years of residence on it and cultivation of it.[20] It should be emphasized that the homestead movement had its origin in the workingmen's party in the East and was taken up by Horace Greeley's *New York Tribune*.[21] Unfortunately, Congress wrote into the measure a provision allowing homesteaders to get title after six months by commuting their claim to a cash entry and paying $1.25 an acre. It was this commutation clause that enabled cattlemen, sheepmen, timbermen, and speculators using dummy entrymen to gain title to land with precious water or rich timber that otherwise would have been unavailable to them. Cash sales were limited to lands that had been offered at public sale. If not purchased at once, they were then opened to private sale. But since public auction sales were largely halted after 1870, it was only through this commutation provision of the Homestead Act, and the Preemption Law and Timber Culture Act, that speculators employing dummy entrymen could acquire large quantities of public land.

Let us look at the status of the public lands in 1880, before any extensive demand for the withdrawal of the lands from purchase or entry was under way. Excluding Alaska, in that year 752 million acres of public land were surveyed and 700 million acres were unsurveyed. The unsurveyed land was as yet unwanted, unsuited for agriculture, and mostly inaccessible.[22] Even the surveyed land that extended all the way from Florida to Washington Territory contained tracts that were not to draw settlers for many years to come. These retained or obviously unwanted lands were not continued in public ownership for any conscious purpose; they continued mostly without supervision or management. But during the intervening years from 1880 to 1912 the acreage of unappropriated and unreserved public domain was reduced rapidly to 314 million acres.[23] By 1913, most of the retained but surveyed land had been open to homesteading for years but was too unpromising to attract homesteaders for the time necessary to acquire a patent. As for the unsurveyed land, it is doubtful whether the staff of the General Land Office thought of it as more than an encumbrance. Even today, there still remain 91 million acres that are unsurveyed, of which the bulk is in Nevada, Montana, Arizona, California, Idaho, and Utah.[24] When an unsurveyed area draws attention because of the discovery of minerals, the surveys can fairly promptly be extended to them.

In the 1880s the commissioners of the General Land Office began to pay more attention to the public lands, especially the surveyed land, and staff was assigned to defend them against timber plunderers without, however, much support from Congress. We must continue to think of the public lands west of the 99th meridian as attracting little attention. If one were a cattleman, why undertake to homestead if stock could be pastured upon the public range without acquiring any obligation or taxes? But stockmen wanting to be sure of adequate grass for their cattle had to try for ownership or, at a later date, to advocate government control of the rangelands.

The "Giveaway System," and the "Great American Barbeque," are terms applied to the widespread corruption in the land administration that occurred as officials connived at extensive transfers of land between 1880 and 1910 by misuse of the Homestead Act, Preemption Act (until 1891), Desert Land Act, the Timber and Stone Act, and Forest Land Scrip legislation.[25] During discussions leading to the making of land grants to railroads, members of Congress had denied that the grant of 20 million acres to the Pacific Railroad and of 45 million to the Northern Pacific would result in land monopoly. They affected to believe that the grants would gradually be broken up like the first grant to the Illinois Central Railroad.

Surely the large sale of timberland in Washington, totaling 1,528,269 acres that the Northern Pacific made to Frederick Weyerhaeuser was contrary to these implied assurances.[26] Yet, who else could have come to the aid of that railroad when it needed to unload its timberlands?

There is one type of statistic that the United States government seems to be loath to compile, as is evidenced by the fact that no data have been collected on multiple ownerships of farmland. The Bureau of Corporations, staffed with Theodore Roosevelt's appointees, examined the ownership of timberland in all parts of the country and published the results in 1913, virtually indicting the large owners of timberland, railroads, companies, and individuals.[27] Its report showed that Frederick Weyerhaeuser and Thomas B. Walker (the largest nonincorporated timber landowners) and others had acquired their lands by purchases from land grant railroads or by buying tracts others had obtained through the use of dummy entrymen and other abuses of the land system. Not that Weyerhaeuser himself or his employees had acquired land illegally, but he had purchased great quantities of land from others who (in the words of William B. Greeley, who later became chief forester) may have used "farcical land procedures."[28] Weyerhaeuser was not responsible for their illegal actions, but his extensive purchases had the effect of boosting prices for timber and enhancing the profits of those willing to use illegal methods. I emphasize the bureau's report because it marks the culmination of efforts to induce Congress to tighten up the land system or at least to slow down the accumulation of large parcels of land, and it was a late, though important, factor in the growth of the conservation movement.

Officials of the General Land Office continued to favor use of the homestead laws to get the public lands into private ownership. To the political spoilsmen there was as yet no thought or intention of keeping the public lands from anyone. Various states were vying with each other to attract immigrants and to counteract the propaganda coming from Canada about richer prairie land that could be more easily acquired and with more help from its government. The General Land Office first began giving out information about the areas where free lands could be acquired in 1913. Soon it was distributing circulars explaining the procedures to be followed in applying for homesteads, listing the acreage available by counties in all the public land states, and roughly characterizing the lands. At the same time, the Reclamation Service was building dams and canals to provide water for irrigating dry land, whether public or private.

Not until Harold Ickes became secretary of the interior in 1933 did that department foster a new spirit of conservation. Under Ickes, the government showed its concern for the retention of the remaining public lands and undertook to improve the public rangelands and manage the forests according to the best scientific practices.

During this time Lewis C. Gray and a group of agricultural economists in the Department of Agriculture's Division of Land Economics were writing about the diminished productivity of hill farms resulting from soil depletion and erosion, the need for resettling occupants of such farms on better land, and the withdrawal of such land from cultivation and its utilization for forests, recreational activities, and wildlife refuges. Gray faulted the Bureau of Reclamation's plans for the irrigation of extensive areas of semiarid land whose crops would need subsidies if they were to compete with imported sugar and rice. He urged the limitation of new reclamation projects to those that had a sound economic basis, but his recommendations fell on deaf ears. Gray wished to prevent the intrusion of homesteaders into the drier rangelands of the West. They were contributing to relief problems and creating the tragedy of the dust bowl.

From these ideas came the submarginal land program of the Resettlement Administration, but Gray was unsuccessful in halting the subsidization of irrigated land that was producing unneeded crops and harming agriculture in other sections. The activities of the Reclamation Service (later the Bureau of Reclamation) indeed have transformed the West but at considerable national expense and not without creating serious national problems.[29] Gray urged that federal money being spent on what he saw as "reckless and ill-planned" reclamation projects should be devoted to taking submarginal lands out of cultivation in the East, Midwest, and South, and resettling the users of these wornout lands elsewhere. Here, it should be emphasized that Gray, Rexford G. Tugwell, and Henry A. Wallace were advocating that the federal government should keep and manage certain types of public land for the common benefit of the United States.

The West wanted no effective curbs on the sale of public lands. In California, where land monopoly had reached a greater degree of concentration than perhaps anywhere else, it was attracting wide attention as a result of Henry George's flaming attacks in his *Our Land and Land Policy* and even more influential *Progress and Poverty*. Sacramento papers and the *San Francisco Chronicle*, in its radical period, took pleasure in publishing the landholdings of prominent speculators, financiers, and lumber kings. Largest of all were the holdings of Miller and Lux and Haggin and Tevis.[30] Some of these large properties have since been acquired by a number of others, while some remain more or less intact.

The movement of corporations into mining, lumbering, steel, and other industries, made possible by their ample credit, resulted in even larger concentrations of land and natural resources in their hands. This fueled the growing antimonopoly movement. Demands for the forfeiture of those railroad land grants of corporations that had not earned them by meeting the conditions of the grants became insistent, and the outcry against large English and Scottish holdings led to the forfeiture of some grants and to the adoption of antialien ownership laws by the federal government and a number of western states.[31]

Increasing attention was given to the plight of would-be settlers searching for free farmland. Agrarian political parties, who advocated limitations on landownership, proposed a tax system that would discourage speculation. It was evident, however, that there was little support for such measures in Congress, and western representatives seemed loath to press for the repeal of laws that were being flagrantly abused to create large ownerships.

Matthew Josephson applied the term *robber barons* to those who had accumulated their wealth by corrupt, inequitable, and antisocial methods, while Theodore Roosevelt called those writers who laid bare their conduct *muckrakers.* These men shared in enlightening Americans about the way capitalists had despoiled their country by stripping off the forest cover, digging up the minerals, leaving an ugly mess and useless land behind, and how they had milked corporations dry, sold dangerous drugs, and built ships, public buildings, and even private houses that were unsafe. One had only to point to the mining regions of Pennsylvania and the cutover regions of the Lake States for some of the worst effects that laissez-faire capitalism had left in its wake.

These attacks upon the robber barons, which began before 1900, by 1930 had led conservationists to insist that parts of the public heritage in lands should be saved from the despoiler, and be put into national parks, national forests, or other forms of public ownership. Many of these muckrakers, influenced by socialism, were moving away from the view that private ownership should be the ultimate goal of public land policy. They were asking whether it was good public policy to allow all the original growth of redwood, Douglas fir, or Sitka spruce to be acquired and cut down by the lumber companies, and whether valuable mineral rights should devolve upon the discoverers of minerals in the public lands no matter what value they might have.

Then, too, there were the enemies of big business. Their vilification was not new, but a high point was reached during the 1880s and throughout the first decade of the twentieth century with Grangers, Green-backers, antimonopolists, Populists, and finally Bull Moose Progressive joining the movement against them.[32] Land grant railroads were attacked because they charged high rates for hauling grain and livestock, for their pooling devices, and for how they exercised their political power in the states. Also, they were criticized for selling off their huge land grants slowly, for separating mineral rights from surface rights, and for their reluctance to pay taxes.[33] Failure to fulfill the statutory conditions of their grants led to demands that Congress declare them forfeited. Some forfeitures were achieved only after long years of litigation.

There had been earlier and unsuccessful attempts to keep valuable natural resources in public ownership. Early in the nineteenth century advocates of conservation had succeeded in having stands of live oak set aside in reserves.[34] This wood was essential for the construction of naval vessels. Notwithstanding, pilfering of the live oak continued, and soon support for the protection of the reserves was lost. An odd provision of the Land Ordinance of 1785 had stipulated that sections 8, 11, 26, and 29 in every township offered at public sale should be withheld "for future sale." Also to be reserved was "one-third part of all gold, silver, lead, and copper mines to be sold or otherwise disposed of as Congress shall hereafter direct." Whatever the meaning of this imprecise stipulation, very little mining was done in the lead district of Wisconsin for some years. In 1807 Congress moved to reserve all lead mines in the district and to lease them for five years. From then until the 1840s mineral land policy was in a muddled state. In time, the lands were sold regardless of their minerals.[35] This failure to manage the lead-bearing lands was to have an important effect on the development of the gold- and silver-bearing lands of California and Nevada. In these states the drafting of rules and regulations and the settlement of disputes about claims were left to the miners to decide. Apart from the protection given in this way to miners' claims, actual ownership of gold-mining tracts was to come late in the day.[36]

Social historians have long been interested in the use of leisure time, as growing numbers of Americans have won shorter hours of labor, longer vacations, and increased incomes that enabled them to take their families traveling. The wealthy had always been able to travel to exotic and historic spots, and mountain and seashore resorts; by 1900, the middle class was also enjoying these delights. The twentieth century was not far along before working people too were on the road, and, as a result, existing parks and camping places were becoming overcrowded.[37] The public began to demand that scenic areas not be marred by destructive cutting of timber, and support escalated for the government's retention of public land for recreation. The railroads must be credited with a share in these developments; they labored hard to develop passenger traffic to Yosemite, the Grand Canyon of the Yellowstone, and Glacier Park.

The first indication of congressional concern for the protection of natural wonders came in 1832 when the mineral springs of Arkansas began attracting visitors to witness their daily flow of millions of gallons of water. Having a constant temperature of 143 degrees, the water was said to provide relief for numerous diseases. To protect the springs from exploitation, four sections of land surrounding the springs were "reserved for future disposal and shall not be entered, located or appropriated for any other purpose whatever." Unfortunately, Congress had waited too long before acting. Numerous claims to land already had been entered on these sections, and were to take a great deal of time and litigation to settle. Visitors came to the reservation for the alleged curative value of the waters, not because of the area's scenic importance. Though it had been designated as a national park in 1921, John Ise thought the reservation was not worthy of that status, and argued that it either be granted to the state or kept merely as a reservation.[38]

Yosemite Valley was the first area of great *scenic* significance to come to the attention of Congress. Tourists recounted stories of its great natural beauty. Heavy depredations were already being made on nearby groves of sequoias, preemption claims had been entered in the Valley of the Merced, and commercialization was already under way in this extraordinarily beautiful area. Uncertain of its role in safeguarding such a spectacular area, Congress, in 1864, dodged the issue by granting the Valley of the Merced to California in the hope that the state would give the area the necessary protection. The grant declared that the valley "shall be inalienable for all time. . . ."[39] As also happened with most of the parks created later, controversy arose over the private claims which had already been made in the valley.

In the meantime, the wonders of Yellowstone were attracting attention, and it was suggested that a park be created there to prevent its glorious wonders from falling into the hands of commercial exploiters. In 1872 Congress enacted legislation mandating that the headwaters of the Yellowstone River were to be "reserved and withdrawn from settlement, occupancy, or sale . . . and dedicated and set apart as a public park or pleasuring ground for the benefit and enjoyment of the people. . . ."[40]

This act marked the beginning of what was to become the National Park System. Since there was no appropriation for the maintenance and protection of the park, about all that the act did was to set aside the

Yellowstone from all forms of private entry but not private use. One writer makes clear that newspapers in nearby areas widely supported the action. Congress had moved from its earlier position that private ownership was best, no matter how wild, how scenic, and how scientifically unique the lands in question were. The next generation moved at a gallop instead of a slow crawl.[41] Today, Congress has established parks and national monuments containing 68 million acres, of which 61 million acres are retained land in Yellowstone, Grand Canyon, Grand Teton, Mount Rainier, Glacier, Olympic, Crater Lake, Sequoia, and Rocky Mountain National Parks and a number of very large parks in Alaska. New parks have also been created by gifts of land to the federal government and in some instances by purchase such as the Redwood National Park, Shenandoah in Virginia, Great Smokies in North Carolina and Tennessee, and Mount Desert in Maine.[42]

Parks were for recreation and enjoyment of nature's wonders and they were open to all. An early objective of the Park Service was to end private ownership of parklands and to bar mining, lumbering, and sheep and cattle grazing, but it was necessary to compromise on all these issues because of rights that had been established before the parks were created.

A more important change in public land policy came when it was feared that the standing timber of the United States was being cut and utilized at such a rate that the supply would soon approach exhaustion. Bulls and bears in the timber market had been jockeying prices back and forth ever since the Census of 1880 had included scandalously low estimates of the amount of timber remaining in Pennsylvania and the Lake States.[43] The result had been to push prices of stumpage high and to bring large profits to the owners of pinelands.

Antimonopoly sentiment, especially with respect to the public lands, had as its principal purpose halting the misuse of the settlement laws and repealing the Timber Culture and Preemption Acts, by placing effective controls in the Desert Land Act and the homestead laws requiring evidence of improvements, thereby making it more difficult to make corrupt use of the commutation clause, and by denying owners of more than 160 acres the right to file on a homestead entry. These reforms were partially achieved in 1891.[44]

Students of the history of forestry and of the lumber industry have been trying to decide how Congress was induced to insert in the Act of 1891 "to repeal the timber culture laws and for other purposes," section 24, that authorized the president to "set apart and reserve . . . in any part of the public lands wholly or in part covered with timber or undergrowth, whether of commercial value or not, as public reserva-

tions. . . ."[45] This section came to be called the Forest Reserve Act.
No legislation marked as sharp a break with the prevailing giveaway
policy as this section 24 of the Act of 1891. Forestry experts at the time
of its enactment, as well as later historians, have marveled that such
unlimited power was given to the president.[46] True, in view of the weak
leadership of presidents since Lincoln, little action might have been
expected, but as soon as the section was used vigorously by Theodore
Roosevelt, Congress was quick to restrict that power. Section 8 of the
act also should be mentioned as it shows how far Congress was willing
to go in defense of persons accused of stealing timber from the public
lands. It provided that "it shall be a defense" if the accused was a resident
of the territory or state and his cutting was for "agricultural, mining,
manufacturing or domestic purposes. . . ." This enabled Anaconda and
Homestead, without fear of prosecution, to cut thousands of cords of
wood on the public lands to shore up the mines.

Since the 1870s, nature lovers and park and forest preservationists
had watched with increasing concern the great American land grab in
which Westerners with little capital took part as hungrily in their small
way as the well-financed cattlemen, sheepmen, lumbermen, and speculators in timberland. By 1890, the superintendent of the census could
say the frontier was gone, though in fact more land was to be taken up
by homesteading during the next two decades than had been preempted
or homesteaded before. But it was time to stop the land grab, especially
for forestlands, rangelands, scenic spots, and desert land capable of
being irrigated. The Act of 1891 had not touched the most abused of
the land laws, the Timber and Stone Act. Moreover, in 1897 and 1899,
Congress enacted measures allowing landowners in national forests and
the Mount Rainier National Park to exchange their barren or sparsely
timbered land within these reservations for richly timbered public land
elsewhere, thereby providing one of the greatest boons to land grant
railroads and other investors in timberland.[47] Nor had the Act of 1891
brought to the administration of the General Land Office more evidence
of integrity. Fifteen forest reserves, containing 12 million acres, were
still being pillaged by lumbermen and had no protection from destructive
fires. Indeed, S. W. Lamoreux, the commissioner of the General Land
Office, said that the forest reserves were given no more protection from
fire and from trespassers than were the unreserved lands.[48]

Efforts to reform the settlement laws had accomplished little, for
advocates of the giveaway were still powerful in Congress and in the
General Land Office. They were able to hasten the transfer of public
lands to private ownership through four acts adopted between 1904 and
1916. In 1904, homesteads of 640 acres were permitted in the semiarid

lands of the sand hills of western Nebraska; in 1909, the Enlarged Homestead Act allowed 320-acre units generally; and finally, in 1916, the Stock Raising Homestead Act permitted homesteads of 640 acres on land suitable for livestock grazing only. Furthermore, in 1912 homesteaders were required to live on their claims only for three years.[49] These acts perpetuated a bad system, wrote Benjamin Hibbard, with the result that "we have a large number of sorry homesteaders; a group of stockmen bankrupt, due in part to the breaking up of the range; a great many acres of good short grass plowed up and rendered worthless for many years to come." Between 1901 and 1910 the greatest of all American land rushes occurred, with 831,640 people, anxious to get their share of the public lands, filing homestead entries averaging 83,184 a year. Many of these were surely legitimate but misguided settlers, though others were filing as dummies for livestock owners.[50]

In the meantime, progress was being made in professional forestry, where conservationism was to be largely centered as a result of the leadership of Carl Schurz, Bernhard Fernow, and Gifford Pinchot. Concern for the rapid destruction of the pineries of the Lake States and the equally rapid disposal of much of the best remaining redwood and Douglas fir land in California and Oregon led Schurz, who was then secretary of the interior (1877–81), to deplore the alienation of large acreages to lumbermen and speculators through misuse of the settlement laws and to provide more effective enforcement of the antipillage regulations. Retention of the timberlands in government ownership would make possible controlled cutting. He urged that timberlands be withdrawn from entry under the homestead and preemption laws and recommended in 1880 that two townships containing some of the greatest of the giant sequoia trees be set aside for the public. His decisions on appeals from subordinate officials were less slanted in behalf of the land grant railroads than had customarily been the case.[51] It was too much to expect that Congress would support his efforts for repeal, but, as Harold Dunham has shown, Schurz did a great deal to draw attention to the plight of the timberlands and the desirability of retaining the best of them in government hands.[52]

Bernhard Fernow, who was trained in forestry in Germany, emigrated to the United States where he quickly became an important figure in forestry education. As an active leader, he wrote extensively in forestry journals, took a major part in forestry organizations, and headed the Department of Agriculture's Division of Forestry until 1898 when he became the head of the Cornell School of Forestry.[53]

After 1898, the subsequent head of the Division of Forestry, Gifford Pinchot, with strong support from President Roosevelt, was the outstanding leader of the conservation movement. Before Pinchot had completed his stint in what had become the National Forest Service, he had increased acreage of the reserves to 160 million acres. A most skillful propagandist for professional forestry, Pinchot made the National Forest Service one of the best administered agencies of the government. He initiated the practice of contracting for the sale of timber to loggers and, in 1905, introduced grazing fees to provide income for improving the rangelands as authorized by amendments to an appropriation act of June 16, 1897 (subsequently termed the Organic Act of the Forest Service). Officials limited the number of cattle and sheep, determined when they would be allowed to graze the tender young grass early in the season and where the sheep would be bedded, provided water, and eliminated noxious plants, as well as plants wasteful of water. Gradually, as care was taken to prevent destructive grazing on steep slopes, the forest ranges improved and the contrast between the condition of grazing lands in the Forest Reserves and those outside became noticeable.[54]

In 1934—in the midst of America's Great Depression, drought, and low livestock prices—Edward T. Taylor, a Colorado rancher, introduced in Congress a measure to set aside or withdraw from entry approximately one-half of the remaining public lands not currently in any reserved status. These lands were to be managed in somewhat the same manner as the rangelands in the Forest Reserves. Subsequently, all the remaining public lands worthy of control were to be placed under organized government management.[55]

The adoption of the Taylor Grazing Act was particularly interesting to critical observers because eighteen years earlier its author had sponsored the 640-acre Stock Raising Homestead Act, which allowed small cattle raisers to homestead on 640 acres of the public lands that had been previously classified as suitable only for stock raising on a small scale. Wilson's secretary of the interior, Franklin K. Lane, with his fondness for homesteading, held that settlement constituted the "best use of the remaining public lands," and hoped it would aid in drawing returned soldiers to the public lands. The measure was badly planned for established stockmen needing more rangelands, and speculators— well aware how rapidly land values were rising—took adantage of it. A 640-acre tract of the deteriorated rangeland was inadequate for a cattle ranch and, as events worked out, much of the land classified as suitable for cattle grazing was used to raise wheat during the high moisture years following the act's passage. The range was broken up, and successive years of drought soon led to the abandonment of much of the land or to its sale to stockmen. One can say only that the Stock

Raising Homestead Act was a grand bust, as Representative Taylor himself admitted in 1934 when he introduced his new bill.[56]

The Taylor Grazing Act marked, one might say, the final step away from the giveaway of the public lands and the beginning of government control of public land not previously part of the national forests. As more investments in improvements were made on the rangelands, "the public's stake grew too, and the concept of retaining rather than disposing of the land took hold."[57]

Meantime, eastern and southern states were anxious to have national forests established within them. They were concerned about the serious damages they suffered from floods after unusually heavy rains. The swift runoff was accelerated by the fact that the timber had been cut at high elevations in the Appalachians and the White Mountains. These states wanted the federal government to acquire some of the lands at high elevations so as to implement modern forestry methods designed to retain water and retard runoff. This proposal won considerable popular support, and in 1911 it was implemented in a modest way when the House of Representatives (by a vote of 130 to 111) and the Senate (by a vote of 57 to 9) passed the Weeks Forest Purchase Act.

This act provided for the creation of the National Forest Reservation Commission to investigate and report on lands in the Appalachians, where modern forest practices might have the effect of reducing damages from floods and improving the navigability of the streams. The land could be acquired only with the approval of the state in which it was located, but once it was, it was "permanently reserved." An amount of $1 million was authorized for such purchases in the first year and $2 million annually thereafter. In 1924 Congress in the Clarke–McNary Act took a further step by allowing purchase of forestland not only on the headwaters of navigable streams but elsewhere "for the production of timber. . . ."[58] That these steps were highly popular is surprising for the older states, like those in the West, looked critically on federal interference in their affairs.

Additions to the Forest Service lands came slowly under these two acts. But during the New Deal, when the Resettlement Administration was purchasing worn-out agricultural lands, many millions of acres in upland areas were rapidly acquired and placed under the administration of the Forest Service, bringing the current total acquisitions of that agency through purchase to 27 million acres in forty-two states, of which 5 million are in the Far West.[59]

Mention also should be made of 8,729,000 acres of land that have been purchased or otherwise acquired for national parks, 6,729,000 acres for fish and wildlife protection, 14,257,000 for defense and use by the

Corps of Engineers for flood protection on major rivers, and 988,000 acres for reconstruction work by the Tennessee Valley Authority in the South. Altogether the United States government has 59,796,000 acres of acquired land and 677,858,000 acres of retained public domain land.[60] It is notable that outside of the twelve public land states in which the government holds large areas there is strong support for these acquisition programs. It is also significant that opposition to the sale of portions of the retained and acquired land is currently being vigorously pressed.

Wayne Aspinall, who became chairman of the House Committee on Interior and Insular Affairs in 1959, strove to have Congress appoint a land commission to deal with the many land problems then facing the committee. Its members were unable to find easy solutions to these land use questions because no organic act existed to guide the Bureau of Land Management nor was there a clear division of authority on which to rely. Aspinall refused to report out measures to meet these needs until the committee would join with him in providing for the establishment of a public land commission with authority and funding to conduct detailed surveys of all such issues. In this way he gained his objective, the establishment of the Public Land Law Review Commission in 1964, with himself as chairman.[61]

Two or three conservationists struggled to tilt the commission's operations slightly from the economic user's point of view but had little success. Chairman Wayne Aspinall was left with virtually complete authority to appoint consultants, hire the staff, and make the decisions. Aspinall recognized that the time was not ripe for an overt attack upon federal ownership of the remaining public lands. Among the 137 recommendations of the commission were proposals for the enactment of an organic act for the Bureau of Land Management, reconsideration of all land withdrawals, federal grants in lieu of the taxes local jurisdictions were prevented from imposing on federal lands within their boundaries, and determination of which Forest Service and Bureau of Land Management lands, if any, could be made available for sale or exchange with the states. The final report, *One-Third of the Nation's Land*, was not altogether satisfactory to the environmentalists or to the economic users, and probably both sides felt the other had won the most. However, the report did say, "We urge reversal of the policy that the United States should dispose of the so-called unappropriated public domain lands."[62] True, it hastened to add that this statement did not mean that all public lands should "remain forever in federal ownership."

The Land Policy and Public Management Act of 1976 carried out some of the PLLRC's recommendations, including the repeal of the Homestead Laws, which many thought should have been repealed not later than 1934. Congress declared that "it is the policy of the United States that the public lands be retained in federal ownership. . . ."[63] Congress weaseled somewhat in the wording of the act, but nevertheless this declaration stood squarely as congressional will. Gov. John V. Evans of Idaho said early in 1979, when the Sagebrush Rebellion was warming up, that the Land Policy and Management Act for the first time declared it is the policy of the United States that "the public lands be retained in federal ownership . . . ."[64] After reading thousands of pages of House and Senate committee hearings on public lands, reports on bills, and discussions in Congress, I am convinced that Aspinall—despite his strong support of the mining interests and his concern for the complete development of the Intermountain West—was a better representative of the West than Secretary of the Interior Watt. This is evidenced by his vigorous support of reclamation projects that were perilously close to being economically questionable, and his awareness that environmentalism had come to stay and that the West must give fair consideration to the views of environmentalists.

Robber barons, in their mad dash for wealth, left millions of acres of unproductive land from Maine to Minnesota. The coal and iron industries, and those searching for gold in the West, left behind piles of slag and debris and abandoned communities, the wealth having gone to Philadelphia, New York, and San Francisco. It would be better, thought the conservationists, to keep the remaining public lands under federal management so that lumbering and mining could be controlled and restoration achieved. To all this western individualists were opposed, but the times favored forms of collective control and preservation of the land.

Public ownership of coal, oil, and gas, owing to withdrawals by Theodore Roosevelt, are bringing large revenues to the government, much of which is promptly shared with the states of origin. In 1980 mineral leases, exclusive of those of the Outer Continental Shelf, yielded $618 million to the United States, the great bulk of which is returned to the states in which these minerals are produced and to the development of irrigation in the arid states. It is not clear to this writer whether any of the states, other than Alaska, has succeeded in developing the minerals on land it has retained or established endowments for their use when the minerals are gone.

Conservation, environmentalism, and ecology all argue that whatever the faults of government management of the national forest lands, the national parks, and monuments, and the bulk of the BLM rangelands, such management is less corrosive of nature and the well-being of man than private ownership has been shown to be elsewhere.

Notes

1. Throughout this paper I have relied heavily upon *The History of Public Land Law Development* (Washington D.C., PLLRC, 1968), which I wrote for the Public Land Law Review Commission, and especially the chapter on mineral lands by Robert W. Swenson. Other useful publications include Roy M. Robbins, *Our Landed Heritage: The Public Domain, 1776–1970* (Lincoln, University of Nebraska Press, 1976); Benjamin H. Hibbard, *History of Public Land Policies* (New York, Macmillan, 1924); and Malcolm Rohrbough, *The Land Office Business: The Settlement and Administration of American Public Lands, 1789–1837* (New York, Oxford University Press, 1968).

2. Henry Steele Commager, ed., *Documents of American History* (7 ed., New York, Appleton–Century Crofts, 1963) pp. 120–121. In a 1890 case involving fraudulent entries of land by a corporation, the U.S. Supreme Court declared that the public lands "were held in trust for all the people" (137 *U.S. Reports*, 170).

3. Commager, *Documents*, pp. 131 and 144.

4. In his great six-volume work, *Jefferson and His Time* (Boston, Mass., Little, Brown, 1948–81), Dumas Malone found occasion to discuss Jefferson's view on public land policies in the first volume only. The invaluable *Jefferson Cyclopedia* (New York, Russell & Russell, 1967), edited by John P. Foley, has many more references to public lands. Unfortunately, the publication of Jefferson's correspondence, edited by Julian P. Boyd, has been long delayed, currently reaching only 1791 some thirty-two years after the first volume appeared. It comprises nineteen volumes without indexes. Temporary indexes for the first twelve volumes have been issued. Hamilton has been treated far more expeditiously, with the twenty-six volumes covering his entire life published between 1961 and 1979, each volume being indexed, and with some very thought-provoking essays included in the notes.

5. *United States Statutes*, I:144.

6. Beverley W. Bond, Jr., "The Foundations of Ohio," in Carl Wittke, ed., *The History of Ohio*, vol. 1 (Columbus, Ohio State Archaeological and Historical Society, 1941) p. 375.

7. Calculated from data in *Historical Statistics of the United States* (Washington, D.C., 1957) p. 712.

8. Act of September 4, 1841, 5 *Stat.*, 456.

9. Paul W. Gates and Robert Swenson, *History of Public Land Law Development*, p. 249.

10. Thomas C. Donaldson, *The Public Domain: Its History With Statistics, House Miscellaneous Documents*, 47 Cong., 2 sess., no. 5, pt. 4, pp. 209–213. After various printings, this widely used document was out of print for many years. A new edition, with an appraisal of its usefulness by Paul W. Gates, has subsequently appeared (New York, Johnson Reprint Corporation, 1970).

11. Harry N. Scheiber, *Ohio Canal Era: A Case Study of Government and the Economy, 1820–1861* (Athens, Ohio University Press, 1969); and George R. Taylor, *Transportation Revolution, 1815–1860* (New York, Rinehart, 1951). Also see Paul W. Gates, "Nationalizing Influence of the Public Domain," in *This Land is Ours: The Acquisition and Disposition of the Public Domain* (Indianapolis, Indiana Historical Society, 1978) pp. 102–126.

12. G. W. Knight, "History and Management of Land Grants for Education in the Northwest Territory," in *Papers*, American Historical Association, vol. 1 (1885); and see also, Hibbard, *A History of Public Land Policies*, pp. 269–288, on the maladministration of the swamplands by the states in their selection and disposal.

13. Acts of March 3, 1875, February 22, 1889, and July 5, 1889, 18 *Stat.*, 475, 25 *Stat.*, 677, and 26 *Stat.*, 216, 222. In 1935 the writer made an effort to compile data showing

the total acres of public lands owned by the forty-eight states and came up with 91,026,223 acres, of which 27,741,272 were owned by nonpublic land states, principally Texas, New York, and Pennsylvania. The largest holdings of the public land states were New Mexico (12,267,045 acres); Arizona (8,894,763 acres); Montana (5,645,887 acres); Colorado (3,463,407 acres); Idaho (3,343,359 acres); Wyoming (3,623,953 acres); South Dakota (2,884,670 acres); Utah (2,841,132 acres); Michigan (2,438,908 acres); Washington (2,352,473 acres); and Minnesota (2,209,452 acres). Eight other public land states still held more than 1 million acres. Some of these holdings, notably in Michigan, were forfeited or tax reverted lands. Worthy of study for comparative purposes are the records of state administration of their lands. The table of state ownership for 1935 is found in *Certain Aspects of Land Problems and Government Land Policies*, in part VII of *The Supplementary Report of the Land Planning Committee to the National Resources Board* (Washington, D.C., 1935) p. 89.

14. I related the story of Miss Dix's efforts to gain public support for the states to build asylums for the care of the insane, and of Morrill's fight for land grants to aid in establishing colleges of agriculture and engineering, in *The Wisconsin Pine Lands of Cornell University* (Ithaca, N.Y., Cornell University Press, 1943).

15. Act of July 2, 1862, 12 *Stat.*, 509; see Paul W. Gates, "Western Opposition to the Agricultural Act," *Indiana Magazine of History* vol. 37 (June 1941) pp. 102–136; and Paul W. Gates, "The Morrill Act and Early Agricultural Science," *Michigan History* vol. 46 (December 1962) pp. 289–302.

16. For the amount of scrip the nonpublic land states received and the acreage of such scrip entered in the western states, see the table in Hibbard, *History of the Public Land Policies*, pp. 333–334.

17. Acts of August 30, 1890, 26 *Stat.*, 418, March 2, 1887 and February 1, 1888, 24 *Stat.*, 441 and 25 *Stat.*, 32.

18. The Newlands Act of June 17, 1902, is found in 32 *Stat.*, 388. The Hatch Act of March 2, 1887, which authorized the appropriation of funds out of the income from public land sales for the establishment of agricultural experiment stations in every state, included no appropriation. On February 1, 1888, Congress made an appropriation for experiment stations through the Department of Agriculture and made no allusion to its coming from the income from public land sales. Thereafter, funds were similarly voted so that no funds from public land sales went to experiment stations. In 1902, with 95 percent of all money derived from public land sales in the fifteen semiarid states going into the Reclamation Fund, there was no certainty that income from such humid states as Arkansas, Alabama, Mississippi, and Minnesota would be sufficient to pay for the agricultural colleges; consequently, any deficiency was met with funds not otherwise appropriated. See Act of June 17, 1902, 32 *Stat.*, 383; and Arthur C. True, *History of Agricultural Education in the United States, 1785–1925* (Washington D.C., Government Printing Office, 1929) pp. 199*ff.*

19. Lawrence B. Lee appraised all major works on reclamation-irrigation in *Reclaiming the American West: An Historiography and Guide* (Santa Barbara, Calif., ABC–Clio, 1980).

20. Act of May 20, 1862, 12 *Stat.*, 392.

21. Helene Sarah Zahler, *Eastern Workingmen and National Land Policy, 1829–1862* (New York, Greenwood Press, 1941).

22. Donaldson, *Public Domain*, p. 189. I have corrected Donaldson's figure for unsurveyed land.

23. General Land Office, *Annual Report* (1912) p. 56.

24. Computed from the Department of the Interior, *Public Land Statistics* (1980) p. 130.

25. William K. Wyant, *Westward in Eden: The Public Lands and the Conservation Movement* (Berkeley, University of California Press, 1982) pp. 32–53, has a useful chapter on "Land Grabbers, the Law, and the Prophets." Compare with Everett Dick, *The Lure*

of the Land: A Social History of the Public Lands from the Articles of Confederation to the New Deal (Lincoln, University of Nebraska Press, 1970).

26. Ross R. Cotroneo, *The History of the Northern Pacific Land Grant, 1900–1952* (New York, Arno Press, 1979), pp. 246*ff.*; and Ralph W. Hidy, Frank E. Hill, and Allan Nevins, *Timber and Men: The Weyerhaeuser Story* (New York, Macmillan, 1963) pp. 207*ff.*

27. U.S. Bureau of Corporations,· *The Lumber Industry*, p. I, "Standing Timber," 1911, p. II, "Concentration of Timber Ownership in Important Selected Regions," 1914, and p. III, "Land Holdings of Large Timber Owners," 1914. Judging by the hundreds of references to Frederick Weyerhaeuser, members of his family, and the Weyerhaeuser Timber Company, it is apparent that Frederick Weyerhaeuser was the colossus of the timber barons.

28. William B. Greeley, *Forests and Men* (Garden City, N.Y., Doubleday, 1951). Greeley, who was chief of the Forest Service from 1920 to 1928, recounts (on pp. 32–38) procedures used to gain ownership of timberlands in the Lake States and the Pacific Northwest, contrary to the purpose of the law. He was well aware of the blatant misuse of the land laws to accumulate large ownerships.

29. This treatment of L. C. Gray is partly based on my recollection of the seminar on Land Policies and Problems at the Brookings Institution in 1933–34, and on my association with him in the Land Policy Section of the Agricultural Adjustment Administration during 1933–35; and it also stems from information found in Richard S. Kirkendall, *Social Scientists and Farm Politics in the Age of Roosevelt* (Columbia, University of Missouri Press, 1966) chap. 1–8.

30. See Paul W. Gates, *Land Policies in Kern County* (Bakersfield, Calif., Kern County Historical Society, 1978), on Miller and Lux and Haggin and Tevis.

31. Paul W. Gates, *Landlords and Tenants on the Prairie Frontier Studies in American Land Policy* (Ithaca, N.Y., Cornell University Press, 1973) pp. 292*ff.*; and W. Turrentine Jackson, *The Enterprising Scot: Investors in the American West After 1873* (Edinburgh, Scotland, Edinburgh University Press, 1968) pp. 102*ff.*

32. Ray Ginger, in *The Age of Excess: The United States from 1877 to 1914* (New York, Macmillan, 1965), reflects the views of Charles A. Beard and Mary R. Beard in *The Rise of American Civilization* (New York, Macmillan, 1927).

33. David M. Ellis, "The Forfeiture of Railroad Land Grants," *Mississippi Valley Historical Review*, vol. 33 (June 1946) pp. 27–60; and "The Homestead Clause in Railroad Land Grants," in David M. Ellis, ed., *The Frontier in American Development* (Ithaca, N.Y., Cornell University Press, 1968) pp. 47–73.

34. See Jenks Cameron, *The Development of Governmental Forest Control in the United States* (Baltimore, Md., Johns Hopkins University Press, 1928), pp. 1–4; and Virginia Steele Wood, *Live Oaking: Southern Timber for Tall Ships* (Boston, Mass., Northeastern University Press, 1981). Wood provides an interesting insight to the dependence on live oak for the construction of early naval vessels.

35. Joseph Schafer, *The Wisconsin Lead Region, Wisconsin Domesday Book* vol. 3 (Madison, Wisc., General Studies, 1932) *passim*; James E. Wright, *The Galena Lead District: Federal Policy and Practice, 1824–1847* (Madison, State Historical Society of Wisconsin for the Department of History, University of Wisconsin, 1966); "Legal Aspects of Mineral Resource Exploitation," in Gates and Swenson, *History of Public Land Law Development*, pp. 699–764.

36. Rodman W. Paul, *California Gold: The Beginning of Mining in the Far West* (Cambridge, Mass., Harvard University Press, 1947).

37. Foster R. Dulles, *America Learns to Play: A History of Popular Recreation, 1607–1940* (New York, P. Smith, 1940).

38. John Ise, *Our National Park Policy* (Baltimore, Md., Johns Hopkins University Press, 1961) pp. 13, 206, 244 *ff.*

39. Act of June 30, 1864, 13 *Stat.* 325.

40. Act of March 1, 1872, 17 *Stat.* 32. In the Senate there was no division on the bill; and the House voted for it 115 to 65. See Aubrey L. Haines, *The Yellowstone Story: A History of Our First National Park*, vol. 1 (Yellowstone National Park, Wyoming, Yellowstone Library and Museum Association, 1977) pp. 156–173.

41. Haines, *The Yellowstone Story*, pp. 165–173.

42. *Public land Statistics*, 1980, pp. 11, 22.

43. For a discussion of the poorly prepared estimates of timber in Pennsylvania and the Lake States for the Census of 1880 that were made by H. C. Putnam, a major dealer in pineland stumpage, and the effect of the low estimates on stumpage prices, see Gates, *Wisconsin Pine Lands of Cornell University*, pp. 115*ff.* See also Sherry H. Olson, *The Depletion Myth: A History of Railroad Use of Timber* (Cambridge, Mass., Harvard University Press, 1971).

44. Paul W. Gates, "The Homestead Act in an Incongruous Land System," in Vernon R. Carstensen, ed., *The Public Lands: Studies in the History of the Public Domain* (Madison, University of Wisconsin Press, 1963) pp. 339–340.

45. Act of March 3, 1891, 26 *Stat.*, 1103.

46. Harold K. Steen, *The U.S. Forest Service A History* (Seattle, University of Washington Press, 1976) p. 26.

47. Acts of June 4, 1897, July 1, 1898. and March 2, 1899, 30 *Stat.*, 36, 621, 994. For treatments of the forest scrip measures see John Ise, *The United States Forest Policy* (New Haven, Conn., Yale University Press, 1924) pp. 176–190 and 292–294; William S. Greever, *Arid Domain: The Santa Fe Railway and Its Western Land Grant* (Stanford, Calif., Stanford University Press, 1954) pp. 57–68.

48. U.S. General Land Office, *Annual Report* 1893, pp. 94–95; and also see Lamoreaux's later report for 1895.

49. Acts of April 28, 1904, February 19, 1909, June 6, 1912, and December 29, 1916, 33 *Stat.*, 547, 35 *Stat.*, 639, 27 *Stat.*, 123, and 39 *Stat.*, 862.

50. Hibbard, *History of Public Land Policies*, p. 401.

51. Joseph Schafer, *Carl Schurz, Militant Liberal* (Evansville, Wisc., Antes Press, 1930), pp. 216–218; *House Executive Documents* 45 Cong., 2 sess., vol. 7, serial 1800, pp. xvi and xix; *House Executive Documents*, 46 Cong., 2 sess., vol. 9, serial 1910, pp. 26–30; *House Executive Documents*, 46 Cong., 3 sess., vol. 9, serial 1959, pp. 33–38.

52. See Harold H. Dunham, *Government Handout: A Study in the Administration of the Public Lands, 1875–1891* (originally published by Edwards Brothers, Ann Arbor, Michigan, 1941, and reprinted by DeCapo Press, New York City, 1970), a work that is too often neglected.

53. Andrew Denny Rodgers, III, *Bernhard Eduard Fernow, A Story of North American Forestry* (Princeton, N.J., Princeton University Press, 1951) p. 112. Rodgers' book contains excellent chapters on the development of professional forestry, its organizations, periodicals, meetings, and political activities.

54. Gifford Pinchot's highly personal account of the growth of professional forestry, the use of government commissions in influencing public opinion in behalf of conservation with the constant support of Theodore Roosevelt, in *Breaking New Ground* (New York, Harcourt, Brace, 1947) is indispensable. Also see, M. Nelson McGreary, *Gifford Pinchot Forester-Politician* (Princeton, N.J., Princeton University Press, 1960); Samuel P. Hays, *Conservation and the Gospel of Efficiency: The Progressive Conservation Movement, 1890–1920* (Cambridge, Mass., Harvard University Press, 1959); and Roderick Nash, *From These Beginnings, A Biographical Approach to American History* (New York, Harper & Row, 1973).

55. Advocates of the final disposal of the grazing lands have made much of a clause in the Taylor Grazing Act that "pending its final disposal" the grazing lands should be

given public management. Ever since, livestock users and writers on land policy have held that this is an indication that the act should subsequently be replaced by measures that would allow the disposal of the lands to private interests. Economic users of the lands have never let this possibility drop from further considerations and they revived it aggressively during the recent Sagebrush Rebellion. Their effort has not gained much support, though in 1978–81 it was widely discussed.

56. Phillip O. Foss, *Politics and Grass. The Administration of Grazing on the Public Domain* (Seattle, University of Washington Press, 1960), p. 33; Steen, *U.S. Forest Service*, pp. 56*ff*; and E. Louise Peffer, *The Closing of the Public Domain: Disposal and Reservation Policies, 1900–50* (Stanford, Calif., Stanford University Press, 1951) pp. 164 and 214*ff*.

57. Bureau of Land Management, *The Nation's Public Lands: A Briefing Package* (Washington, D.C., 1980) p. 15.

58. *Congressional Record*, June 24, 1910, 61 Cong., 2 sess., p. 9027, and 61 Cong., 3 sess., February 15, 1911, p. 2602; Act of March 11, 1911, 36 *Stat.*, 963; Act of June 7, 1924, 32 *Stat.*, 652.

On the background of the Weeks Act, see two articles by Charles D. Smith, "The Mountain Lover Mourns. Origins of the Movement for a White Mountain National Forest, 1880–1903," *New England Quarterly* vol. 33 (March 1960) pp. 37–56; and "The Appalachian National Park Movement, 1885–1901," *North Carolina Historical Review* vol. 37 (January 1960) pp. 38–65.

See William E. Shands and Robert G. Healy, *The Lands Nobody Wanted, Policy for National Forests in the Eastern United States: A Conservation Foundation Report* (Washington, D.C., The Conservation Foundation, 1977). The report deals with the eastern national forests. The title is scarcely apt, for, unlike millions of acres of unsurveyed public lands for which there was no demand for two centuries, much of the land acquired under the Weeks Act had long since been in private hands and in some areas they were increasing in value as a result of the desire of urban residents to gain a site for a second home, if no more than sufficient for a small cabin.

59. *Public Land Statistics*, 1980, p. 13.

60. In the nineteenth century efforts to assure to the government part of the returns from mining were very largely failures, producing more litigation than income. The reservation of minerals other than coal and iron in the later land grants to railroads, for example, is the occasion of recent discoverers of minerals other than coal and iron to bring action against the original owners of the grants (railroads) declaring that all such minerals at this late date are still owned by the government and subject to entry, though the title was passed to other owners a century or more ago. The Stock Raising Homestead Act of December 28, 1916, which provided for 640-acre stock-raising homesteads on land classified as fit only for grazing, contained a reservation "of all coal and other minerals in the lands . . . together with the right to prospect for, mine and remove the same. . . ." 39 *Stat.*, 864. Under this measure, 33,181,000 acres went to patent on which the federal government held mineral rights. When the vast coal resources of the Great Plains states began to attract investor's attention, the possibility of mining coal on cattle and wheat ranches in Montana, the Dakotas, and elsewhere produced much excitement by owners fearing their best lands might be seriously damaged by mining. Similarly, the Northern Pacific railroad had reserved mineral rights on its land. President Hoover proposed to complicate title questions even more by urging that the grazing lands be conveyed to the state but with mineral rights reserved by the United States.

61. Act of September 19, 1964, 78 *Stat.*, p. 13.

62. *One-Third of the Nation's Land. A Report to the President and to the Congress by the Public Land Law Review Commission* (Washington, D.C., PLLRC, 1970) p. 1.

63. Act of October 21, 1976, 90 *Stat.*, 2743.

64. *Rangeland Policies for the Future*, Proceedings of a Symposium held on January 28–31, Tucson, Arizona (Washington, D.C., Government Printing Office, 1979) pp. 10–12.

3

Why Have We Retained the Federal Lands?

An Alternative Hypothesis

BARNEY DOWDLE

Anyone with even a casual interest in the history of the public lands is well aware of the work of Professor Paul W. Gates. He has spent more than fifty years studying, reflecting, and writing on the subject and has participated in the development of some of the more important milestones in the evolution of public land policy.

As a relative newcomer to the field, I was indeed honored to have been invited to comment on Professor Gates's chapter, "The Federal Lands: Why We Retained Them." Somewhat presumptuously, I have titled my comments "Why Have We Retained the Federal Lands?" and have added the subtitle "An Alternative Hypothesis." This implies that my interpretation of the issues which Professor Gates has addressed differs from his. This implication is correct.

I am certain this is not a new experience for Professor Gates. History in general, and public land history in particular, has been characterized by debates over conflicting hypotheses. The so-called safety valve theory of migration and land settlement,[1] and Frederick Jackson Turner's hypothesis regarding the role of the "frontier" in shaping American institutions and culture,[2] are good cases in point. Moreover, I am sure that Professor Gates appreciates that important truths can emerge from debates which are precipitated by different interpretations of history.

I should preface my remarks by noting that my interests have been largely confined to the subject of government ownership of commercial timberlands; the reasons for their existence, and the economic and political implications of their management. This is merely a subset of the public land issues that Professor Gates has considered. He has, I might note, discussed the historic practices of the lumber industry, and the background to the creation of the national forests at some length in his chapter. We do, therefore, have common interests.

If there is a justification for inviting me to comment on Professor Gates's chapter, perhaps it can be found in the Greek proverb about the hedgehog and the fox. "The fox knows many things," it is said, "but the hedgehog knows one big thing."

If I may portray myself as a hedgehog, then the "one big thing" that I think I know—and which differs from what Professor Gates has reported to be his belief—concerns the rationale and necessity of government timber production. Professor Gates's views on this issue, as reflected in chapter 2 and in his monumental *History of Public Land Law Development*, fall into the category of what I have labeled the "market failure" hypothesis.[3]

The Market Failure Hypothesis

Briefly, the U.S. government got into the business of timber production in the United States on a large scale at the turn of the century. This occurred primarily because it was believed that the market system would not work in this important area of production—profit seekers would liquidate naturally endowed stands of mature timber; that is, that they would "mine" them, but subsequently would not grow timber as a crop on these cutover lands. Continued reliance on the market system, therefore, would lead eventually to a timber "famine." Because of the importance of timber and wood products, as well as the many other benefits which forests produced, this was considered to be an intolerable situation.

Students of public land history are aware that—largely as a result of the acceptance and implementation of the market failure hypothesis—about 27 percent of the nation's nearly 500 million acres of commercial timberlands are presently in some form of government ownership. Additionally, since past timber harvests have been concentrated on private lands, private timber inventories now consist mostly of immature timber stands.

Government timberlands now contain more than 60 percent of the total inventory of mature softwood sawtimber. The lumber and plywood industries are dependent upon this inventory. Having a timber-processing industry dependent on government timber, which in turn is managed largely on the basis of noneconomic criteria, has serious economic implications. Yet they do not receive the attention that they deserve. Considerable evidence has accumulated suggesting that this is a highly unstable and wasteful relationship.

In my opinion, the market failure hypothesis, which is implicit in Professor Gates's chapter, and which he appears to endorse, is incorrect. It is not consistent with generally accepted economic theory nor with all relevant facts, a conclusion which I will develop below.

A fundamental and profound consequence of accepting the market failure hypothesis and changing land-disposition policies so that public domain timberlands are no longer alienated into private ownership is that the nation has acquired a dual economic system for timber production. This system is both private and public. Privately owned timberlands provide a basis for the market, or "free enterprise," system to guide timber production decisions, and I will characterize the public sector as "socialistic." Webster defines *socialism* as ". . . governmental ownership and administration of the means of production of goods," and I feel that this definition can be applied to public timberlands as well.

Expectedly, the management of the public, or "socialized," sector has been politicized from its beginning. Numerous vested interests—including the timber, mining, and grazing industries, and, somewhat belatedly, recreational users and environmentalists—have organized to influence management in order to benefit themselves at the expense of the general public.

Forestry schools in the United States also have a sizable vested interest in publicly owned timber management. Most of them originally were established to train the foresters ("planners") who would manage public timberlands. Prior to World War II, graduates of U.S. forestry schools largely were employed by federal and state timber-management agencies. Forestry schools today are steeped in the antimarket beliefs that helped to bring government into the business of timber production.

Arguments—or more correctly, assertions—which commonly state that nonmarketed multiple-use benefits ("public goods") from the forest were a major reason for changing public land policy and establishing public forests are not convincing. The historic record, including implementing legislation and administrative regulations, frequently refers to relationships between forest cover and benefits other than timber, especially water flows. For the most part, however, these observations were used primarily to embellish the argument that the market system did not work in timber production.

At the risk of appearing cavalier in dismissing a highly important issue, it would be difficult, I think, to justify the existing system of publicly owned timberlands as being necessary for the production of public goods. Parks, wilderness areas, and other exclusive-use areas are separable issues, and a stronger case can be made for government ownership of lands which are dedicated to these uses. The question which needs addressing is, What kinds of institutional arrangements perform best, if judged by relevant performance criteria, for producing optimal mixes and amounts of marketable timber and nonmarketable multiple-use benefits? Which is better—regulated private ownership, or public ownership and management? A more complete discussion of these questions would be useful, but is beyond the scope of the present review.

"Socializing" the West

Given the political climate which existed near the turn of the century, when policy was changed to retain rather than dispose of public domain timberlands, it is not surprising that vast acreages of grazing and other nontimbered lands would also remain in government ownership. The historic sequence of land disposal, from east to west, and the timing of this shift resulted in most lands under government ownership being located in the West.

More than half the total land area of a number of "public land" states in the West is in federal ownership. Disregarding Alaska, Nevada leads the list with more than 86 percent of its total land area owned by the federal government. Two government agencies were subsequently established, the Forest Service and the Bureau of Land Management (BLM), which manage this vast land area.

Professor Gates is obviously aware of the distinction between public and private economic activity—that is, capitalism and socialism—and he mentions both economic systems in his chapter. The Forest Service and the BLM, which centrally plan production and distribution of many services for lands under their jurisdiction, meet the definition of socialist institutions.

Much of Professor Gates's discussion of capitalism is not very complimentary. Populist "villains" such as "monopolists," "big business," despoilers," "looters," and "speculators" are all mentioned. He observes, more or less approvingly, that the "muckrakers" were disillusioned with "the way capitalists had despoiled [resources]," and notes that "one had only to point to the mining regions of Pennsylvania and the cutover regions of the Lake States for some of the worst effects that laissez-faire capitalism had left in its wake."

Professor Gates also observes that "many of these muckrakers, influenced by socialism, were moving away from the view that private ownership should be the ultimate goal of public land policy." More simply stated, they preferred socialism to capitalism. It is my opinion that these views obviously prevailed, as is evident in the extensive U.S. system of public land ownership, the management of which is centrally planned. Professor Gates appears to accept, if not endorse, this "socialistic" system as being economically and socially desirable. Prior to elaborating my own views on how this situation developed, a synopsis of public land history, as interpreted by Professor Gates, is in order.

The Substance of Public Land History

Much of public land history deals with such topics as how the United States acquired land, why and how it alienated this land into private ownership, the politics and problems associated with the disposal process, and how policies changed as this process took place. Much of Professor Gates's chapter is devoted to these topics, and this is a story that he obviously tells very well.

The philosophical architects, such as Thomas Jefferson and Alexander Hamilton, are discussed, along with congressional and private sector power brokers, relevant laws, and how these laws fared in their implementation. There are no surprises in Professor Gates's discussion. He is going over familiar ground, and the substance of his explanation is accepted by most authors of history textbooks used to educate American students.

The privatization of public domain lands was not a controversial issue in the early days of the new Republic. The Founding Fathers strongly believed in private property rights and, by implication, in an economic system which was based on private ownership of land resources. Private property rights were associated with individual freedom, which was of considerably greater concern than economic efficiency in the allocation of resources. The adjectives, "free enterprise" or "capitalistic," which would be used today to label such a system, did not come into popular use until a much later date.

Numerous arguments did arise, however, over the best means of getting land into private ownership. Should public domain land be used to provide pensions for war veterans, or sold, or bartered for infrastructure such as schools, canals, and railroads, or used as an inducement for land settlement? In fact, each of these methods was employed at one time or another for establishing private property rights.

As Professor Gates clearly indicates, the privatization process produced a lot of graft and corruption, and public domain lands were frequent targets of thieves and looters. I might note that the magnitude of this theft is not usually estimated quantitatively by public land historians, and, in my opinion, it may have been assigned more importance than it deserves.

Most of the resources stolen from the public domain were not worth very much. As an example, the distinguished American historian, Louis M. Hacker, has reported that the value of lumber products sold in the United States in 1870 was $168 million.[4] Assuming that all of these lumber products were produced from stolen timber, once production and freight costs are deducted the thieves could not have gained very much.

The argument can also be made that much of this theft occurred in violation of bad laws and regulations. Housing was obviously more important to society at that time than standing timber. One is reminded of Adam Smith's (1776) observation regarding smugglers: "A person who, though no doubt highly blameable for violating the laws of his country, is frequently incapable of violating those of natural justice, and would have been, in every respect, an excellent citizen, had not the laws of his country made that a crime which nature never meant to be so."[5]

Eventually, doubts did arise that the kind of economic system which the nation was building was adequate to some of the tasks that it was being assigned. Some of these doubts had their origins in ideological objections to capitalism, or laissez-faire, as the system was called. Their substance was that the system was not "fair," income and wealth were not being "properly" distributed.

Doubts also arose among staunch supporters of the existing economic system because the facts being generated by the process of settlement and economic development, and "best available interpretations" of these facts, raised serious questions about its effectiveness. Some resources were being mistreated and squandered, for example, buffalo and beaver. Timber, which will be discussed below, was also frequently mentioned.

Some of the questions which arose during the nineteenth century remain with us today. What is the best way to manage resources such as air and water, and fish and wildlife? How can we assure the preservation of topographical features of the landscape and species of flora and fauna which are in some sense unique?

It should be emphasized that, from the standpoint of economic and legal analysis, we understand many of these questions much better today than we did at the time that institutional arrangements were being developed for the management of these resources. Public ownership and management, and regulations which apply to private ownership, have not always performed as well as was initially assumed. In many cases, the use of exchangeable property rights and the price system have proved to be a better way of developing incentives for using resources efficiently. (The feasibility of using price incentives for controlling pollution, as opposed to regulatory standards, is discussed in the 1982 *Economic Report of the President.*[6])

Once established, however, institutional arrangements quickly give rise to vested interests. These vested interests can become formidable obstacles to the application of improved analytical skills and knowledge in developing desirable institutional reforms. A good case in point is the current opposition of hunters in the West to the sale of public lands. Recipients of subsidies usually are not enthusiastic about discussing their subsidies. Nor is it considered good form to subject them to analysis and public debate.

After several years of experience in the field of public land management, I have gained a good appreciation of the difficulties of initiating a disinterested, constructive dialogue on these issues. "Political realities" are continually getting in the way.

The Economics of Building an Economic System

In describing events which were encountered along the route to the privatization of public domain lands, Professor Gates is a good tour guide. He is less helpful in explaining the implications of arriving at a different destination than was originally intended. Perhaps a different analogy might help our understanding.

The development of an economic system based on private property rights on a largely unsettled continent, such as that which occurred in the United States for more than a century, can be viewed as a massive building project. If this analogy is used, then public land history can be interpreted quite differently than it has been by Professor Gates.

Cost overruns, graft, corruption, and theft are commonly associated with government construction projects, especially those which are undertaken on a "crash" basis. Clearly, the privatization of public domain lands during the latter half of the nineteenth century can be described in this manner.

During the period between 1875–1900 an average of 22,000 final entries were filed annually with the General Land Office under the provisions of the Homestead Act of 1862. Annual average land disposals for these entries was 3 million acres,[7] an area about the size of the state of Connecticut. If land grants for other purposes are added, then it is obvious that a frantic scramble for land was occurring. Congress not only encouraged this scramble, but also was highly permissive in accommodating it.

That many government activities associated with the privatization of the public domain were costly and morally reprehensible, given existing conditions, is not very surprising. Under the less frantic conditions of the present, government housing, highway, water, and energy projects are beset with similar problems. The Washington Public Power Supply System (WPPSS) in the Pacific Northwest, which has recently defaulted on its bonded indebtedness, hardly commends itself as an example of efficient government planning.

Once a government project has been completed, it may be no less useful because of cost overruns encountered in its construction. Overly costly highways may still improve the flow of traffic. More serious problems arise if a project does not work, or is left uncompleted. The WPPSS, with its abandoned nuclear projects, has had to face this embarrassment.

With these thoughts in mind, the following question can be raised about building an economic system, Should an economic system be condemned merely because it has been costly to build? In my opinion, it is not obvious that it should, especially in view of the fact that in the United States many of these costs appear to be a result of bad decisions by government. In his classic 1936 article on the Homestead Act, Professor Gates has discussed a number of relevant examples.[8]

The historic record is replete with examples of socialist economic systems which have been much more costly to build than the U.S. free enterprise system. The Soviet Union and the Peoples Republic of China provide excellent examples. More important, in spite of their high "construction costs," these economic systems do not work very well. Individual freedom and economic efficiency do matter, and the record is reasonably clear that free enterprise has outperformed socialism in achieving both these objectives.

If traditional explanations for the retention of U.S. western public lands, which are used primarily for commercial production, are reinterpreted in terms of costs associated with building a free enterprise economic system, then at least two categories of construction costs can be identified, both of which are consistent with economic reasoning and a considerable amount of factual evidence. The first includes costs associated with the privatization of nontimbered lands under the provisions of the Homestead Act. A second category includes costs associated with the privatization of public domain timberlands. Both categories require some further explanation.

There Is No Such Thing as "Free" Land

Economists are inclined to observe that "there's no such a thing as a free lunch." The substance of this aphorism is that lunches may be free to some people at some time, but eventually the recipients, or others, must pay for them. Economists also observe that if market-clearing prices are not charged for resources and goods and services which are scarce, and subject to exchange, then other means will be found for clearing the market. Queuing is a good example of this: if an asking price is less than the market equilibrium price, then people will stand in line.

If the Homestead Act is viewed in the perspective of these generally accepted economic propositions, then one is led to conclude that the land was not free. If the land had value, and if a zero price was being charged, then a disequilibrium condition existed in the land market and other means would have been found to reach an equilibrium. What were these other means?

Professor Gates sheds some light on this question. As he observes, "Congress, in 1862 . . . promised any citizen, or prospective citizen, title to a *free* homestead of 160 acres of surveyed land after *five years of residence on it and cultivation of it*." (Italics added for emphasis.)

Land which was acquired subject to these conditions, especially on the "frontier," required at least some immediate investment expenditures, and initial cash flows were generally negative. Many settlers endured considerable deprivation while waiting for their "free" land to acquire positive value. Not uncommonly, it did not. All these costs can be interpreted as the "purchase price" of land which was acquired under the provisions of the Homestead Act.

When the price paid by settlers was too high, it was to be expected that they would become disillusioned with, and react against, the economic system that "free" land was supposedly helping to build. This was especially true in the Great Plains states, where the art of dryland farming was poorly understood. Manifestations of this reaction appear, for example, in the platforms of the Populist and Progressive parties.[9]

The political response to these developments was to stop privatizing public domain lands. More important, the nation discontinued the extension of a market economy in important areas of resources management. The baby got thrown out with the bath water. In my opinion, a socialist system of land management was established as a rival alternative. Today the grazing lands which are managed by the BLM constitute a large part of this system.

"Cut-and-Run" Timber Exploitation

Professor Gates has correctly observed that timber was the first important resource area in which public land policies underwent fundamental change. Alienation was discontinued, and a policy of retention was adopted. The reasons for this fundamental shift in policy are well known. Professor Gates, as well as other public land historians, have generally accepted the migratory practices of the lumber industry during the nineteenth century as evidence of market failure.

In my opinion, it is more reasonable to hypothesize that migration was a one-time development cost which was unavoidable because of an overabundance of naturally endowed timber inventories. If the market failure hypothesis is untenable, then what is the correct interpretation of the migratory practices of the lumber industry? Simply stated, the market was having a clearance sale on naturally endowed timber inventories. It was behaving as expected, given the structure of timber prices and production costs which existed at the time.

Timber will not be grown in a market economy if it is cheaper to buy. Plentiful supplies of "virgin" timber were available in the United States until the early 1950s, and timber prices remained below replacement (growing) costs until that time. Expectedly, lumber production shifted toward the harvesting of low-priced virgin timber. The industry was migratory.

Lumbermen would not have survived had they attempted to comply with the admonishments of foresters and conservationists to grow second-growth timber crops, which would have provided a raw materials base for a permanently located lumber industry.

The social costs of a migratory lumber industry were undeniably high, but this experience is not unique in U.S. history. Agriculture, mining, railroads, and textiles have also engendered high social costs as these industries were transformed by economic growth. The lumber industry is unique, however, because its migratory practices led to the collectivization of a significant amount of the nation's timber production.

This generally accepted misinterpretation of the migratory practices of the lumber industry was encouraged by a myth which had long existed within the forestry profession. This myth was that the market system would not work in timber production because of the long time required to grow a timber crop. Carl Schurz, Bernhard Fernow, and Gifford Pinchot—all mentioned by Professor Gates—were among its chief perpetrators.

The substance of this myth is that private investors have no incentive to plant and grow timber because the maturation period of their investments is likely to exceed their life expectancy. Apart from altruism, who is willing to bear the cost of planting a tree if he or she does not expect to be alive to realize the gains from its harvest?

The answer to this question, which also dispels the myth, is that anyone should be willing to plant a tree who can reasonably expect to get a return on his or her investment. Moreover, one does not have to wait until the tree reaches maturity, and is harvested. If markets exist for goods-in-process, so that the old can sell their assets to the young, then the market system and profit-seeking will ensure that timber is grown.

The development of private "tree farming" in the United States, especially in the South where most timberlands are privately owned, provides ample and convincing evidence to support this conclusion. In fact, the wood-processing industry has been migrating to the South in recent years despite the fact that most of the nation's softwood timber industry is located in the West; although most of the western inventory is in government ownership. The promise of stable communities with which government forestry began is not being realized in practice.

Conclusions

Professor Gates has amassed a great amount of descriptive detail during the course of his lengthy career in public land history, much of which is summarized in chapter 2. Moreover, this detail is invaluable as a source of information for interpreting the past. As is well known to anyone who attempts to analyze history and other subjects, the tedious, day-to-day task of gathering empirical data is highly important.

If an effective analysis is to be made, however, then theory must also be used to serve as a guide for raising questions that ought to be asked, and for developing hypotheses which are specific enough to be tested empirically. Without a rigorous and coherent theory to serve as a guide, it is easy to become immersed in descriptive detail, and to lose sight of what really may be occurring.

To the extent that theory exists in Professor Gates's works, then it must be inferred. In my opinion, he begins his analysis of the building of our economic system with an implicit theory of capitalism; land was to be privatized. He concludes, however, with an implicit theory of socialism; land would be retained in public ownership. Building costs encountered along the way—which, understandably, were poorly understood, and which many found morally reprehensible—appear to have caused this shift.

My point is that theory—in history it might be more appropriate to speak of orthodoxy—which is being used to guide analysis, whether used explicitly or implicitly, should not be permitted to go unquestioned for prolonged periods of time. The reason that it should not is that theory determines which questions get asked, and—what is perhaps more important—those which do not.

By neglecting theory we can compound confusion over time. If ongoing research leads to the accumulation of evidence which is inconsistent with previously accepted historic interpretations, then what should be done? If we use timber as a case in point, then it is correct to observe, I think, that government became involved in timber production near the turn of the century because of the belief that this was an area of market failure. At present, the market system convincingly demonstrates that it is an efficient system for producing timber, and in fact is doing so more efficiently than is government.

Professor Gates and other public land historians may correctly chronicle the evolution of facts. To the best of my limited knowledge, however, they have yet to integrate all relevant facts into their analyses and conclusions about the viability of the market system in timber production. Do markets operate differently today than they did in the past? Or, is it possible that public land historians are guilty of faulty analysis and have, therefore, permitted the untenable market failure hypothesis to be perpetuated?

Notes

1. Roy M. Robbins, *Our Landed Heritage: The Public Domain, 1776–1936* (Lincoln, University of Nebraska Press, Bison Book, 1942).

2. George Rogers Taylor, ed., *The Turner Thesis: Concerning the Role of the Frontier in American History* (Lexington, Mass., D. C. Heath, 1949).

3. Paul M. Gates, *History of Public Land Law Development,* prepared for the Public Land Law Review Commission (Washington, D.C., Government Printing Office, 1968).

4. Louis M. Hacker, *The Triumph of American Capitalism* (New York, Simon and Schuster, 1940).

5. Adam Smith, *The Wealth of Nations* (New York, Modern Library, 1937) p. 849.

6. Council of Economic Advisers, *Economic Report of the President* (Washington, D.C., Government Printing Office, 1982).

7. Benjamin Horace Hibbard, *A History of the Public Land Policies* (Madison, University of Wisconsin Press, 1965).

8. Paul M. Gates, "The Homestead Act in an Incongruous Land System," in Vernon Carstensen, *The Public Lands: Studies in the History of the Public Domain* (Madison, University of Wisconsin Press, 1936) pp. 315–348.

9. Richard Hofstader, *The Age of Reform* (New York, Vintage Books, 1955).

4

The Federal Lands Today
Uses and Limits

PERRY R. HAGENSTEIN

The federal lands today are vast but limited in the contributions they are likely to make to society. The character and potential uses of these lands have been determined by their history. Limits are imposed not only by the nature of the lands and the resources they contain or by their location, but also by the body of laws, policies, and traditions that have grown up about them.

The purpose of this chapter is to describe briefly the present extent and uses of the federal lands and to comment on some matters that are important to their current and prospective status and use. The focus of this volume on the rationale for continued federal ownership of these lands clearly has affected the choice of data and my interpretation of them. It is not my intent here, however, to make a case for or against continued federal ownership.

Marion Clawson, in his new book, and Robert Nelson of the Department of the Interior, in some unpublished papers, provide more detailed information on the federal lands and their uses.[1] Much of the information here is based on these reports, but this chapter should not be considered a substitute for these more detailed sources.

Here we are concerned with the broadly similar onshore lands, outside of those in Alaska and Hawaii, that are administered by the Forest Service and the Bureau of Land Management (BLM). Other federal lands are extensive and important in many ways, and information on

74

their area and use is often lumped with that from the Forest Service and BLM lands.

The lands managed by the Forest Service are the national forests, some of which were reserved from the public domain, while others were acquired by purchase or other means. The BLM lands are primarily the unappropriated and unreserved remnants of the original public domain. These—together with certain other lands and interests in property managed by the BLM, but excluding the Outer Continental Shelf (OCS)—were designated as the "public lands" by the Federal Lands Policy and Management Act (FLPMA) of 1976.

The administration of the federal lands is continually evolving. One reference point I have used used is 1970, the year in which the report of the Public Land Law Review Commisson (PLLRC) was published following an exhaustive review of federal land laws and policies. All of the issues discussed at this workshop were touched on to some extent in the PLLRC report. Although much has happened since 1970 in the continuing debate over uses of the federal lands, this marks but one brief period in the overall evolution of federal land policies. I believe the frequent references to the PLLRC report offer an appropriate perspective to the current debates. The issues have been addressed before.

Restrictive arrangements that limit what can be done to change the status and uses of the federal lands include a wide range of customs and practices that have grown up around federal land administration. Many have become embedded in law, but others—which are just as effective in limiting what can be done with these lands—remain as customs. They could be thought of as tacit covenants.

On the whole, they favor local interests and users. Any proposals for changes in the status of the federal lands must come to terms with the various and complex arrangements that now apply to their use. Proposed changes, which may seem simple at first glance, will inevitably run into resistance from those who benefit directly or indirectly from the present arrangements.

Three important changes have occurred since 1970, and each limits changes in the status and uses of the federal lands:

● Land use planning is now required by law for the national forests and public lands.

● Payments were increased to state and local governments in recognition of the tax immunity of federal property.

● Various environmental protection restrictions were adopted, which have generally been applied more readily and with greater impact on the federal lands than on other lands.

The Federal Lands

Currently the federal lands include some 705 million acres of surface onshore lands and an estimated 560 million acres of seafloor on the Outer Continental Shelf (OCS) to a water depth of 200 meters. If the area of the OCS is measured to water depths where current technology permits development of oil and gas resources, the total area of the OCS is perhaps two or three times as great. In addition, the federal government holds about 66 million acres of reserved mineral interests in lands whose surface is privately owned. All minerals are reserved on some 42 million acres, coal only on 16 million acres, and other specified minerals or combinations on another 8 million acres (table 4-1).[2]

The total acreage of federal ownership has not changed much in the past two or three decades, other than in Alaska. There, grants of federal land to the state under the Statehood Act and to natives in Alaska under the Alaska Native Claims Settlement Act have reduced federal ownership by about 80 million acres over the past twenty years. When the statehood and native grants are satisfied, an additional 68 million acres will have gone out of federal ownership.

The total area is enormous, but the sheer extent of the federal lands suggests a degree of control over the nation's resources that simply does not exist. In addition to limits in the physical ability of the federal lands

TABLE 4-1. ESTIMATED AREA OF FEDERAL LAND, 1982
(in thousands of acres)

Agency	Total	Alaska only	Eleven western states[a]
Forest Service	190,719	23,611	139,403
Fish and Wildlife Service	85,105	76,059	5,564
Bureau of Land Management	317,087	141,800[b]	174,867
Park Service	70,797	51,015	14,084
Defense Department	30,327	1,924	16,846
All other	10,982	303	5,763
Total federal	705,017	294,712[b]	356,527
Total area (excluding inland water)	2,271,343	365,481	752,948

Source: Area estimates for the eleven western states and for the nonpublic land agencies are from Bureau of Land Management, *Public Land Statistics, 1980* (Washington, D.C., Government Printing Office). Area estimates for the public land agencies in Alaska and total areas are for 1982.

[a] Arizona, California, Colorado, Idaho, Montana, Nevada, New Mexico, Oregon, Utah, Washington, and Wyoming.

[b] Of the BLM area, 67.7 million acres remain to be transferred to the state of Alaska and Alaska natives.

to support resource uses, law and practice constrain what can be done with much of this area. In the following section, I will examine the extent to which the area of federal lands available for general resource use is reduced by allocations of land to particular purposes.

National forests and public lands

The Forest Service is responsible for nearly 191 million acres of land, of which 139 million acres are in the eleven western public land states and about 23 million acres are in Alaska. Most of the land was reserved from the public domain, but some has been acquired, mainly in the eastern states and there are some differences in the authorities for managing reserved and acquired national forest lands. The main management thrust for the national forests, both those reserved from the public domain and those that were acquired, is not limited to particular uses. Special land designations within the national forests, however, restrict some uses in some places.

The BLM is responsible for managing a land base that has shrunk substantially in the last two decades. In 1970 it was 470 million acres, but by 1979 it was down to 398 million acres, of which 222 million acres were in Alaska and 175 million acres were in the eleven western public land states. The area in Alaska was reduced by about 100 million acres in 1980 when the Alaska National Interest Lands Conservation Act reserved lands for parks, refuges, and national forests. The area of BLM lands in Alaska will be further reduced to about 74 million acres when assignments of federal lands are finally completed under the Alaska Native Claims Settlement Act and the Alaska Statehood Act. In addition, something less than 500,000 acres of indemnity grants to states other than Alaska remain unsatisfied. About half this area remains to be claimed by Utah. This is also a claim on the BLM lands that will ultimately have to be satisfied.

Of the total area of public lands, about 2.1 million acres of revested O&C Railroad grant lands and 75,000 acres of reconveyed Coos Bay Wagon Road lands in western Oregon are administered by the BLM, but under different statutes than those that apply to the bulk of BLM lands. (About 460,000 acres of Oregon and California grant lands are also managed by the Forest Service, but with special provisions for sharing revenues with local governments in Oregon that are similar to those that apply to the other O&C lands.) Also, as noted earlier, the BLM administers the 66 million acres of reserved mineral interests where the surface lands are privately owned.

Both the Forest Service and BLM have responsibilities in addition to land management. The Forest Service has a major forestry research program and also promotes forestry on nonfederal lands. The BLM is responsible for leasing minerals on all federal lands, including the national forests and the OCS, and for administering the mining of hardrock minerals. It also surveys federal lands, including those administered by other agencies, maintains land records, and is responsible for issuing certain kinds of rights-of-way across federal lands.

Parks and refuges

Two other major systems of federal lands are the National Park System, administered by the National Park Service, and the National Wildlife Refuge System, administered by the Fish and Wildlife Service. These lands, too, are quite varied in character and use.

The National Park Service is responsible for national parks, monuments, historic sites, and some national recreation areas. It manages about 71 million acres of land, of which 51 million acres are in Alaska, 14 million acres in the eleven western public land states, and nearly 6 million acres in the eastern United States. The purposes of management generally are to protect important aspects of our natural and cultural heritage and to provide recreation opportunities. The majority of the lands administered by the National Park Service were reserved from the public domain.

As part of a broad program for preserving, protecting, and enhancing fish and wildlife, the Fish and Wildlife Service administers some 85 million acres of federal lands in wildlife refuges and ranges. Nearly 90 percent of this, about 76 million acres, is in Alaska and over 90 percent of the total was reserved from the public domain. The major purposes of protecting and enhancing wildlife permit other uses to be made of wildlife refuges where they will not preclude achieving the major purposes. Thus, the land can be managed in some cases to permit timber harvesting, domestic livestock grazing, recreation, and even mineral development.

Miscellaneous lands

The Outer Continental Shelf is the federal estate in the seabed extending outward from the 3-mile limit (or 3 marine leagues in the case of Florida and Texas). The 200 meters of water used as one way of defining the outer limit of the OCS is physically precise, but is less meaningful with respect to development of resources. Rights to develop resources of the seabed extend well beyond this limit and are limited primarily by the availability of suitable technology. The major use of the OCS has been for production of oil and gas, but there is potential for the extraction of other minerals in the future.

The importance of the OCS for its energy resources is evident. Production of oil and gas from the OCS rose rapidly during the 1960s, but has risen only slowly since then. About 9 percent of the nation's oil and over 20 percent of its natural gas production now comes from the OCS. Federal revenues from leasing oil and gas development rights on the OCS in recent years have been about ten times those from leasing all minerals, including coal and phosphate as well as oil and gas, on federal onshore lands.[3] In part, this is because resource values on the OCS are high. But, it has also been due to the fact that all OCS leasing is competitive, while most federal leasing of oil and gas onshore is not.

Lands administered by the Department of Defense include about 30 million acres, of which some 17 million are in the eleven western public land states, about 2 million in Alaska, and over 11 million in the states generally east of the 100th meridian. The Defense Department lands include about 8 million acres administered by the Corps of Engineers, some of which are used for recreation. In addition, domestic livestock grazing, timber harvesting, and hunting, fishing, and other forms of outdoor recreation take place on some of these lands.

Finally, there are about 11 million acres of other miscellaneous onshore federal lands. More than half this area is administered by the Bureau of Reclamation and over 10 percent by the Energy Research and Development Administration, mainly in sites used by what was formerly known as the Atomic Energy Commission. The rest includes post office sites, Coast Guard facilities, and a host of other administrative uses.

The total area of the miscellaneous lands is significant. In some cases, there are important issues concerning the use and status of the lands. As with the parks and refuges, however, these issues are usually specific to a particular situation. There seems to be little argument over the overall purposes served by federal ownership and management of these lands.

Alaska: A special case

Each public land state has some unique features that beg for attention in setting federal land policies. Alaska was often singled out for special treatment in the past because of its relative lack of a developed economic base and the fact that almost all of its land was federally owned. In some ways, Alaska has become more like the other states over the past ten years. Nevertheless, there are some important reasons why Alaska should not be lumped with the other states in this discussion of the federal lands.

The most important reason is that the statehood grant selection and confirmation process is still proceeding. The Alaska statehood grants of 104.6 million acres, which were to have been selected by 1984, were the largest ever, whether measured by gross area or percentage of the state's area. The transfer of less than half of the total area has been confirmed. Thus, the federal lands situation in Alaska is in flux and will be for some time. The BLM's responsibilities for lands in Alaska are being sharply reduced, in part because of the statehood grants and partly because of the designation of parks and refuges in the Alaska National Interest Lands Conservation Act of 1980.

Beyond this, the role of the BLM lands in Alaska, especially over the next few decades, has to be viewed as uncertain. As the state selections are completed and the grants are confirmed, the state will be in a position to determine and guide its own future. As the Public Land Law Review Commission (PLLRC) noted, "Haphazard disposals of [federal] public lands . . . could be contrary to the well developed plans made by the state for regional and local land use, and could burden the state and local governments with additional responsibility without corresponding benefit."[4] Vast areas of Alaska have no developed transportation system or other social infrastructure. Disposal of federal lands and development of federal resources could force the state to invest prematurely in such infrastructure and interfere with the state's plans for its own lands. Until the statehood grants are completed and there is a clearer perspective for viewing use of the BLM lands, little should be said about their use.

The national forests in Alaska have long been subject to special policies that were conceived as being in the best interests of the state. Timber from the national forests must undergo initial processing in Alaska before it can be exported from the state, and long-term timber sales have been tied to the construction of major processing plants in Alaska. One important result of this has been that the timber industry in Alaska, which is largely dependent on the national forests, primarily supplies forest products to Japan. Thus, it is not really a part of the U.S. domestic timber economy.

Dedicated lands

Three categories of dedicated federal lands—wilderness areas, wild and scenic rivers, and national trails—are superimposed on lands administered by the Forest Service and BLM, as well as on lands administered by the National Park Service and Fish and Wildlife Service. While these dedicated lands are managed by the agency with general responsibility for the surrounding area, the limitations placed on the uses of the dedicated lands remove them from many of the uses that can be made of other federal lands.

Lands in all three categories are designated by Congress. The effort that has gone into designating each area suggests to me, at least, that their boundaries, as well as the limitations on some uses in them, will not change soon. As indicated by their names, the three designated categories each concentrate on specific recreation uses, although these areas are by no means the only areas of Forest Service and BLM lands that provide these recreational opportunities.

Designated wilderness areas encompass by far the largest area of the three categories in the contiguous states. There are now 25.4 million acres of designated wilderness on the national forests, of which 5.4 million acres are in Alaska and 1.4 million acres—including the Boundary Waters Canoe Area in northern Minnesota—are in the eastern states. The area of wild and scenic rivers on the national forests is about 250,000 acres. Only a small area of wilderness has been designated on the public lands so far, but 24.3 million acres on the public lands and 12.4 million acres on the national forests in the contiguous states have been defined as "study areas." Actions inconsistent with the possible designation of these study areas as wilderness are prohibited until Congress decides whether each area should be made wilderness.

Congress is still in the process of adding to the National Wilderness Areas Preservation System, which, like the National Park System and the National Forest System, may never be complete. It is unlikely that all of the study areas will be designated as wilderness. A significant part of the proposed areas will, however, ultimately become part of the wilderness system.

Other kinds of designations that limit the possible uses of federal lands are national recreation areas, which include some acres of national forest lands, and cultural and archeological areas on both national forests and public lands. Recreation is the primary, although not necessarily the only, use of the national recreation areas. In addition, the Endangered Species Act makes protection of listed species of plants and animals the primary criterion for deciding on uses wherever (and whenever) these species are found to be present.

Summary

When the various subtractions from the total area of federal lands have been made, the area relevant to the remainder of this chapter shrinks from a total of 705 million acres of federal land to about 285 million acres, not including the reserved mineral interests held by the federal government. These are the national forests and public lands in the contiguous states, less the 20 million acres of designated wilderness and 37 million acres in wilderness study areas, that are sometimes described as the federal "multiple-use" lands (table 4-2). They are the lands where practically all of the commodity uses and much of the recreation use take place.

These remaining federal lands vary greatly in character and capability, but two generalizations are useful. First, the federal lands are not as highly productive for most commodity uses as are lands in private ownership. Most of them remained in federal ownership because they were passed over for settlement or resource development under the laws and conditions of fifty years or more ago. Even the national forests that were reserved from the public domain just prior to and following the turn of the century were created after the most accessible forestland had gone into private ownership. The public lands, in particular, have been passed over by generations of potential farmers, ranchers, and miners who succeeded in getting patent to the more productive lands.

John Black once noted that there are no submarginal lands, only submarginal uses.[5] He pointed out that land can be used in an economically efficient manner for some purpose if it can be assembled in appropriate tracts. Thus, some federal lands are economically manageable for timber, some for grazing livestock, and some for recreation, as long as they can be assembled and used for these purposes. Generally, this is the way in which the federal lands are managed and used. It takes

TABLE 4-2. ESTIMATED AREA OF MULTIPLE-USE LANDS IN THE FORTY-EIGHT CONTIGUOUS STATES, 1980
(in thousands of acres)

Surface use	National forests		Public lands[a]	Total
	Eastern	Western		
Commercial forest	20,786	61,404	4,141	86,331
Grazing[b]	407	36,659	149,514	186,580
Miscellaneous[c]	1,402	26,205	21,633	49,240
Total	23,995	142,868	175,288	342,151

Source: Estimates for 1980 based on U.S. Forest Service, An Assessment of the Forest and Range Land Situation in the United States (Washington, D.C., Government Printing Office, 1980); U.S. Department of the Interior, Public Land Statistics, 1980 (Washington, D.C., Government Printing Office, 1980); and agency estimates of area of designated wilderness, which are not included in totals.
[a]Includes 421,000 acres in the eastern states.
[b]Does not include grazing that occurs on commercial forestland.
[c]Includes noncommercial forestland, barren land and mountain tops, and administrative sites.

fairly large tracts if semiarid federal rangelands are to be used efficiently for grazing or if low-site forestlands are to be used efficiently for timber. The various kinds of federal lands are managed in tracts appropriate to their uses.

There are some exceptions to the generally low quality of federal lands. The western Oregon forestlands managed by the BLM and some national forest lands are good timber-growing lands. It also is likely that there are some valuable undiscovered mineral deposits on the federal lands. Even here, prospectors for both energy and hardrock minerals have combed these lands thoroughly for years. It is difficult to find minerals, and changing technology and economic conditions in some cases will undoubtedly make what are now judged to be low-grade mineral deposits economically attractive in the future. But overall, the use of the federal lands for commodity production is closer to the margin of economic suitability than private lands.

The federal lands may be more significant nationally for their contribution to noncommodity uses. Some of the federal lands not already designated primarily for recreation use, such as parks and wilderness areas, provide comparable recreation opportunities. More important, they provide a range of opportunities for recreation that use extensive areas of relatively wild land but do not require truly primitive surroundings, for example, big-game hunting.

Second, most of the federal lands with characteristics that *now* would qualify them as national parks or national wildlife refuges have been reserved for these purposes. Passage of the Alaska National Interest Lands Conservation Act in 1980 marked a major expansion of the two systems. But as interests in the federal lands change, there will certainly be some candidates for parks or wildlife refuges among the present national forest and public lands, especially from some of the wilderness areas. The federal lands still provide a reservoir of lands for meeting park and refuge needs.

In sum, the national forests and public lands in the contiguous states are extensive. In some cases, they are very productive for particular uses. But their extent and potential contribution to satisfying a range of wants can be overstated. They are limited by designations that remove part of them from some uses and by their character and productivity.

Uses of the Federal Lands

The national forests and the public lands are the multiple-use lands, although uses generally are separate in space or time. Considering the major uses separately does little violence to understanding the ways in which they are managed. In fact, each use is typically managed under its own set of policies, which were originally established to meet particular conditions of time and place. Consistency in the treatment afforded the various uses has never been a hallmark of federal land policies.

Each of the major uses of the national forests and public lands in the contiguous states is described in the sections that follow. My purpose is to present a basic picture of the use of these lands, but as with the description of the lands themselves, this discussion of uses is sketchy.

Timber

Timber is the most important commodity resource of the federal lands in terms of its national role. More than one-quarter of the nation's softwood sawtimber harvest comes from the federal lands each year. Most of the forestland on the federal lands is, appropriately, part of the national forests. But the national forests are not entirely forestland, nor is all of the federal timber found on the national forests. The public lands include some forestland, primarily of relatively low productivity. The BLM also is responsible, however, for the management of the O&C Railroad grant lands, the Coos Bay Wagon Road grant lands, and some public domain lands in western Oregon, all of which are very productive for timber.

The 82.2 million acres of commercial forestland (land that is capable of producing at least 20 cubic feet of timber per year and on which timber harvests are not prohibited) on the national forests in the contiguous states amounts to exactly one-half of the total national forest acreage. The remainder of the national forests is grassland or high mountain areas, or is forestland reserved from timber harvesting. Of the commercial forestland on the national forests, 33 percent is of low productivity; that is, it is capable of producing only 20 to 50 cubic feet of wood per acre each year. Another 39 percent is capable of producing 50 to 85 cubic feet per year. The national forests have nearly 20 percent of the nation's high-quality timberland (more than 120 cubic feet per acre per year). But this is a relatively small area, since land in this quality class only makes up 10 percent of the nation's total commercial forestland.[6]

The lowest of these productivity classes (20 to 50 cubic feet) is generally not suited to long-term timber production and harvesting. In many cases, the timber on the low-productivity lands is valuable and could be harvested economically. However, prompt regeneration is required by law on national forest lands. When the cost of regeneration and other costs of harvesting and management are considered, the value of the timber on some of the low-productivity lands is not sufficient to cover all of the associated costs.[7]

The rules that require regeneration of low-productivity lands are based on concern for protection of the land and have little immediate economic justification. In addition, Forest Service decisions to allow timber harvesting on sites where the value of the timber is not sufficient to cover road-building costs and the costs of regeneration similarly have little economic justification.

Timber harvests from about 2.4 million acres of BLM lands in western Oregon were about 12 percent of the total harvest from federal lands in the contiguous states and accounted for about 23 percent of the total value of the federal timber harvest in 1980.[8] This harvest is from an area that is equal to about 3 percent of the commercial forest area on national forests in the contiguous states, which reflects the fact that the lands are more productive than the average national forest lands.

Management of timber resources on the national forests and public lands epitomizes the approach of what Culhane calls "progressive conservation."[9] The resource is visible, measurable, and consistent from year to year. Further, if timber is not cut this year, it stores well on the stump and can be cut next year. Both the Forest Service and BLM target a timber-harvest level for commercial forestland annually or by decade. This target level is based on biological considerations and accounts for the various factors that reduce the target level below the biological potential. A quantity of timber that is as close to the target level as appropriations allow is offered for sale each year. The timber is allocated among users generally with competitive sales procedures. These procedures for managing timber resources provide a model of "rational management" that cannot be readily copied in the management of the other federal land resources.

The actual level of timber sales each year usually falls somewhat below the target harvest levels overall, although there are areas where sales are at or near the target levels. Each sale must usually be harvested within a three- to five-year period. Actual timber harvests from federal lands in any year, which depend on market conditions, may vary considerably from the level of timber sales. As markets for forest products fluctuate from year to year, annual harvests rise above and fall below target harvest levels.

The average volume of timber sold from the national forests and public lands annually has been nearly constant since the mid-1960s.[10] But the national forests' share of total harvests in the United States fell from

1962 to 1976. Even for softwood sawtimber, of which the national forests have slightly more than 50 percent of the nation's total, the national forests' share of the total U.S. harvest fell from about 28 percent in 1962 to 23 percent in 1976. For comparison, softwood sawtimber harvests from forest industry lands, which have 22 percent less commercial forestland than the national forests, increased from 33 percent in 1962 to 38 percent of the nation's harvest in 1976.

Some of the fall in the national forests' share of the U.S. timber harvest is probably due to wilderness designations and other losses of the commercial forestland base on the national forests. There was a decrease of 8 percent in the commercial forestland area on the national forests from 1962 to 1977. But other constraints on federal timber harvests, especially those imposed by the way in which target harvest levels are set, have a greater impact on harvests. The even-flow harvest rule, which restricts timber harvests in a period to a level that at least can be maintained in the future, limits the extent to which national forest harvest can respond to secular increases in total national timber harvests.

Grazing

Grazing of domestic livestock, along with recreation, is one of the most widespread uses of the national forests and public lands in the contiguous states. Almost all of it occurs in the West. Although extensive, grazing on federal lands is generally light when compared with grazing on improved pastureland. It is also often intermingled with other uses, especially with timber production on the national forests and with wildlife wherever it occurs.

The importance of grazing on federal lands arises from the fact that it is the only commodity use on large sections of these lands. In addition, it is integrally tied to, first, the control over water and, second, the use of private farm- and pasturelands. Spring and summer grazing on federal lands must usually be combined with forage from private lands during the winter months. This has brought about the system of "commensurate rights" that ties grazing on federal lands to ownership or control of sufficient private forage to support the livestock when they are not on federal lands.

Total authorized grazing use on the national forests and public lands together decreased from about 22 million animal-unit-months (AUMs) in 1960 to 18 million in 1980.[11] All of the decrease was on BLM lands, reflecting the increasing recognition of other values and uses of these lands. The BLM lands now support slightly fewer units of grazing than do the national forests. The gradual decline in grazing use of the federal lands has been coincident with a gradual increase in total livestock production in the United States. In 1970 the Public Land Law Review Commission estimated that the federal lands accounted for about 3 percent of the forage used by livestock.[12] Although grazing is locally important and the cow–calf operations of the West still supply animals for feedlots, grazing on the federal lands is not nationally important.

Grazing is, however, an important use when viewed in terms of the degree of support it provides for economic activity in broad areas of the western United States. This accounts for much of the apparent influence that grazing and the western livestock industry have had on federal land policies. The extent to which the livestock industry and other user groups have been able to dominate federal land policies has been considerably lessened with the broadening of the economic base of many of these areas and with the rise in numbers and importance of other interest groups.[13]

As with timber, grazing levels are set to be more or less compatible with the biological capability of the land to support continued grazing. Information from the Forest Service and BLM indicates that about 21 percent of the range on federal lands is in the highest of four condition classes based on the condition of the site's vegetation and soil relative to its potential. Only about 11 percent of the nonfederal range is in this best condition class.[14] It is not clear that the standards used for measuring the conditions were exactly the same for federal and other rangelands. But the comparisons do suggest that (1) the condition of federal rangelands is at least equal to that of other rangelands, and (2) the condition of nonfederal rangelands is low.

Targets for the level of grazing, like targets for timber harvest, have been established for each area of federal lands. Unlike timber, however, range forage is an annual crop that does not carry over from year to year. Also, actual range conditions vary more from year to year than targets for grazing use because of variations in rainfall.

This variability in the condition of the range from year to year, perhaps more than any other single factor, has led to a continuing battle between those whose livestock graze on federal lands and the federal managers, whose job it is to assure that reasonable levels of range productivity are maintained. For timber, the sustained-yield rule uses regulation of harvest levels (actually, sales levels) to assure that the overall ability of the land to sustain continued timber harvests is not impaired. But timber stores well on the stump, whereas range forage has to be used, if at all, when it grows. To assure that the federal range receives reasonable protection in years of scant rainfall, average grazing levels have to be well below levels that would be possible in better than average years.

Thus, ranchers can point to forage that goes to waste in some years because permitted grazing levels are lower than what the range would sustain in those years. The dilemma is not readily amenable to the "rational procedures" used in managing federal timber.

Minerals

Minerals development on the federal lands has long been both important and at the root of contentious issues. The issues are as complex as the finding and development of the minerals themselves. The fact that the quantity and economic accessibility of the minerals is not known until considerable effort has been spent on their development is basic to any consideration of mineral policies on federal lands. Equally important is the fact that deposits range in value from high to negative. This fact, which is less frequently mentioned by mining interests than the previous one, is basic to an understanding of the relationship between mineral development and other uses of the federal lands.

Federal reserved mineral interests under private lands deserve mention before we focus attention on minerals on national forests and public lands. Development of these minerals can pose serious problems where it interferes with the desires of the surface owner. For those federal minerals that are leased (for example, oil and gas), the problems are perhaps no more serious than in cases where private mineral and surface rights are divided. That is, issues can be resolved as the leases are being negotiated. For hardrock minerals open to prospecting under the location and claim system of the 1872 Mining Law, problems have arisen where prospectors have begun to establish claims to minerals underlying private property. The prospector has a right to explore for federal hardrock minerals whether they underlie federal or private lands.

Parallel problems arise where mineral rights under federal surface rights are privately owned. This is a common situation on acquired national forests in the Appalachians where much of the coal is in private hands. Obviously, conflicts between surface and subsurface uses are resolved most readily when the same party owns both mineral and surface rights.

Federal reserved mineral interests will continue to raise problems as long as the minerals are potentially valuable and to the extent that they are open to unfettered prospecting. The Public Land Law Review Commission recommended that the federal government retain these reserved mineral interests, but that clearer guidelines for resolving conflicts with surface uses be developed.[15] In the Surface Mining Control and Reclamation Act of 1977, Congress gave surface owners what amounts to a veto over mining federally owned coal beneath their lands. But nothing has been done to resolve potential conflicts where the federal government has reserved title to other minerals.

Leasable minerals on the federal lands are those found mainly in sedimentary deposits—oil and gas, coal, phosphate, potash—and all minerals on acquired federal lands. The leasing process allows the same kind of resource management approach as is used for timber. That is, judgments can be made about the number of leases to be issued in any period, the specific areas to be leased, and the conditions under which development of the minerals will be allowed relative to other uses of the land. In practice, oil and gas leases generally are made available when and where there is a potential lessee. Coal, on the other hand, is now leased on the basis of management plans.

Oil and natural gas production from federal onshore lands are each about 5 percent of the U.S. total annual production. Over 100 million acres of the federal lands (that is, about 40 percent of the total multiple-use area) are under lease for oil and gas. Although the total acreage under coal leases is only about 1 million acres, coal production from the federal lands is now about 10 percent of the U.S. total and the share is increasing. Thus, production of these energy minerals from the national forests and public lands contributes significantly to the nation's total. In addition, these lands produce substantial quantities of uranium, phosphate rock, and potash.

Hardrock minerals on the public domain federal lands are developed under the location system spelled out in the 1872 Mining Law. On acquired federal lands, hardrock minerals are leased in a manner similar to oil and gas. The location system for hardrock minerals permits mining on claims that are still in federal ownership and requires only a nominal fee for patenting claims. No data are collected on the quantity or value of hardrock minerals taken from federal lands. There is no doubt, however, about the important place that the federal lands have played in the development of these minerals. Most hardrock mines in the western states are on private lands that were originally located and claimed as mineral-bearing lands when they were in federal ownership.[16]

The location system assigns a *de facto* priority to hardrock minerals relative to other uses on federal lands that have not been withdrawn from mining. In practice, the leasing system for oil and gas, and perhaps also for coal, is administered in a way that assigns priority to these energy minerals. Thus, there is presumption in law for hardrock minerals and in practice for energy minerals that, where they occur on federal lands, they should be developed more or less regardless of other values on the same lands.

Development of both locatable and leasable minerals on the federal lands, aside from those that have been specifically withdrawn from mineral entry, presumably has progressed as rapidly as it would have if these lands were not in federal ownership. In fact, the absence of royalty or other payments to the federal government for locatable minerals is an inducement for more rapid development on federal lands than on other lands.

Where mineral values are high, they may be many times as high as values attributable to surface uses. In such cases, automatically assigning minerals a priority over other uses may make good sense. But as with all other uses of the federal lands, mineral values vary widely from place to place. Where minerals exist in quantities that can be physically mined, but their value is low relative to surface values, assigning automatic priority to minerals development makes little economic sense. The fact that the quantity of minerals on many claims has been insufficient to meet the "marketability" tests required for patenting claims shows that mining, too, can be a submarginal use of federal lands.

Outdoor recreation

The federal lands provide a wide range of opportunities for outdoor recreation. Together with the national parks and wildlife refuges, they provide most of the nation's opportunities for extensive backcountry recreation in relatively wild surroundings. The national forests, especially, also provide a substantial share of the opportunities throughout the country for downhill skiing, an intensive use of land for recreation.

There seems to be little doubt that total recreation use of the federal lands has been increasing, but the relationship between the level of recreation use and the area of federal lands available and suitable for particular uses is not clear. A recent study for the Pacific Northwest showed that use rates for most kinds of dispersed recreation on the national forests in that region, not including wildlife appreciation, increased when roads were built into previously inaccessible areas, as one would expect.[17] Extending this result to argue that overall recreation use of the federal lands could be increased by changes in land use, however, seems tenuous, at least in the absence of increases in population.

The extensive and varied recreation uses of federal lands tend to be segregated among uses and through time. Much of the extensive backcountry use takes place in wilderness or other designated areas where recreationists have sought and obtained some degree of protection for their particular use. But even more recreation use takes place on the multiple-use lands, where administrative restrictions are sometimes imposed on other uses in order to protect recreation-related values.

There is no available summary of the extent of the area of federal lands affected by administrative designations to protect recreation uses or the degree to which such limitations impose on other uses. One Forest Service study of several western national forests concluded that the effects on timber production of administrative restrictions to protect recreation and related values are substantial and outweigh efforts to increase timber production through more intensive management.[18]

Values that can be attributed to recreation and other uses of the federal lands vary from place to place to the extent that comparisons based on aggregate data can be misleading. The Pacific Northwest study mentioned above concluded that timber values outweigh the incremental recreation values achieved by keeping roadless national forest lands inaccessible by car.[19] There was sufficient variation in these comparisons of aggregate data for major subregions in the Northwest, however, that would not warrant the blanket conclusion that timber values in this region always exceed the incremental recreation values of keeping roadless areas inaccessible by car. Obviously, similar comparisons in other regions of the country might give quite different results.

Watershed protection and wildlife

Protection of watersheds to prevent downstream flooding was one of the purposes for which national forests were established and managed. The importance of maintaining forest cover in preventing floods was disputed during the debates over the 1911 Weeks Law, which provides for the establishment of national forests in the eastern states.[20] Nevertheless, watershed protection has remained an important management objective on the national forests. In the semiarid West, national forests cover much of the high-rainfall zones in the mountains. Managing forests to make more water available for downstream users is now seen as having some importance for areas with overall limited water supplies.

On the more arid grazing lands, watershed protection is one of the purposes of keeping a suitable vegetative cover. With the much smaller amounts of water falling on these lands, management to enhance downstream flows is not as effective as it is on forested lands. Similarly, there is relatively little that can be done with land management to influence water flows on higher-elevation, but unforested, federal lands.

Wildlife management on the federal lands deserves mention because it is recognized as an important use in law and in practice. Because of their extent and character, the federal lands provide critical habitat for some species and they provide extensive opportunities for hunting and viewing. At the same time, it must be recognized that other lands also support wildlife and wildlife-oriented recreation.

Commentary

As "leftovers," the federal lands are not highly productive when compared with those lands that passed from the public domain into private ownership. Yet, they contribute some 25 percent of the nation's annual timber harvest, about 5 percent of its oil and natural gas, some 2 to 3 percent of the nation's annual forage production for domestic livestock, and extensive opportunities for outdoor recreation of kinds that are not readily available on other lands. In addition, they are available for prospecting for needed hardrock minerals.

Practically all of the federal lands are used for one or another of these purposes. Over time, the arrangements for private uses of these lands have been adjusted to reflect economic realities. Thus, land is made available for grazing livestock or for timber harvesting in ways that complement use of private lands and in ways that reflect market requirements. On much of the land where particular commodity uses would be submarginal under other conditions, low prices—or what amounts to user subsidies—make commodity uses economically attractive.

Conflicts among uses of the federal lands are heightened by disparities in prices. The lack of charges for recreation and hardrock minerals and the low prices charged for grazing undoubtedly encourage use of the federal lands for these purposes. The federal lands, despite their status as leftovers, are used and have value for various uses. Since the uses are private, they presumably are consistent with economic efficiency criteria, but they are clearly affected by the various adjustments to economic conditions that are imposed by the hodgepodge of policies for federal lands.

There is no obvious critical national role now played by the multiple-use lands of the national forests and public lands. Where particular uses have required some protection, it has usually been forthcoming. National parks, wildlife refuges, and wilderness areas have been established. Congress has declined to set priorities among the general uses on the remaining lands, other than for that assigned to hardrock minerals. Thus, priorities are set case-by-case and largely on the basis of meeting local needs. As values and needs change, the inevitable resistance to changing established uses has created most of the conflicts that are a familiar part of federal land policies.

The multiple-use lands contain, or are, reserves for meeting increased demands for various uses. There are currently unutilized and undiscovered mineral and energy resources. The national forests have a very large timber inventory, which could be drawn upon in times of national need. The level of outdoor recreation could be increased substantially without greatly impinging on other uses of the lands, and the annual growth of timber and forage for domestic livestock and the production of fish and wildlife could be increased somewhat with increased management inputs. Some of the lands could also be used for agriculture and more intensive residential or commercial uses under some circumstances.

But to note that the federal lands could be used differently, perhaps more intensively, does not distinguish them from most private lands. It is true that they have a sizable part of the nation's timber inventory and many opportunities for more extensive outdoor recreation. The level of most of these uses, including timber and recreation, is set, however, by market forces in relation to comparable uses on private lands. Federal land resources, again with the exception of timber, have not been held as reserves against unexpected events. In fact, low prices for some of these uses have probably encouraged more rapid development than if they had been in private hands.

Receipts from federal land resources

Annual receipts from the sale of federal land resources and for the use of federal lands are in the same range, say, as annual sales of a single large forest products or mining firm. They have increased rapidly in recent years due primarily to increases in prices for energy minerals and timber that have been greater than general price inflation (table 4-3). Gross receipts for the national forests and public lands in 1980 were about $2.3 billion. (If receipts from the leasing of oil and gas on the OCS were included, the total would be $6.4 billion.[21]) Of the $2.3 billion, receipts from minerals leasing accounted for about 40 percent and receipts from sale of timber, including the value of purchaser-constructed roads and payments for regeneration of harvested stands, for about 50 percent.

Policies for charging for federal land resources vary from one resource to the next. Timber is sold in competitive auction. Only a small part of the onshore oil and gas leases, those in known producing geological structures, is sold competitively. The remainder are allocated on a first-come, first-served, basis or by lottery for lands where previous leases have lapsed. In either case, lessees pay a royalty fixed as a percentage

TABLE 4-3. ESTIMATED USES AND RECEIPTS FOR MULTIPLE-USE LANDS, FOR 1970 AND 1980

Use	National forests		Public lands	
	1970	1980	1970	1980
Timber				
Volume sold, million board-feet	13,382	11,399	1,662	1,121
Value, in millions of $	$317[a]	$1,968[a]	$82	$366
Grazing				
Permitted use, in 1,000 AUM	9,285	9,757	10,981	8,875
Fees, in millions of $				
	$4	$18[b]	$5	$20
Minerals				
Oil and gas leases, in 1,000 acres	[c]	[c]	61,334[c]	100,117[c]
Estimated royalties, in millions of $			$91	$605
Coal leases, in 1,000 acres	[c]	[c]	1,497[c]	979[c]
Estimated royalties, in millions of $			$1	$32
Recreation				
Visitor-days, in millions	173	234	39	70

Source: Adapted from Marion Clawson, *The Federal lands Revisited* (Washington, D.C., Resources for the Future, 1983).

[a]The value of timber cut on the national forests was $308 million in 1970 and $730 million in 1980.

[b]Estimate based on reported use and estimated average grazing fee.

[c]Mineral data are reported for all public domain land, which includes most western national forests. They do not include data for mineral leases on acquired lands, which include most eastern national forest lands. In 1980 receipts for oil and gas leases on acquired lands were 5.2 percent of those on public domain lands, and receipts from coal leases on acquired lands were less than 0.5 percent of those on public domain lands.

of the wellhead value of the oil and gas. Coal leases are also sold competitively, when they are made available. As noted earlier, there is no charge for hardrock minerals when production comes from claims on federal lands and the charge for patenting hardrock claims is miniscule.

Rights to graze livestock are now (since 1978 and until 1985) leased at rates that are set by a statutory formula tied to livestock prices. Grazing fees have always been well below the going price for leased grazing on private lands.[22] There is no charge for dispersed recreation use of the federal lands, but charges are levied for the use of campgrounds and similar facilities. Operators of ski areas pay fees that are intended to capture the fair-market value of the permit and they, in turn, charge for use of ski facilities.

Each pricing system has its supporters, but few interests seem concerned about the apparent inequities between resource users. The Public Land Law Review Commission recommended that there should be a uniform basis for pricing goods and services from the federal lands and that fair-market value should be the standard for all such goods and services.[23] Nothing has been done since these recommendations were made to provide greater uniformity.

The federal lands are a large enough source of recreation and timber, and perhaps some minerals, that policies for setting prices, or the way in which the amount put on the market is determined, can affect prices in private markets. The availability of free dispersed recreation on large areas of federal lands, especially in the West, effectively precludes charges for use of private lands for comparable recreation.

For timber, the quantities made available to the market each year are less than what would be made available by a private owner in a competitive market, and this affects prices in private timber markets. A recent study for the Pacific Northwest concluded that prices for standing timber on all ownerships in western Oregon and western Washington would fall sharply below forecast prices if public timber were made available to the market in response to economic demand and supply factors.[24] That is, if the federal government acted with respect to its large timber inventory in the Pacific Northwest, as we would expect a private owner to act, the inventory would be reduced at a rate consistent with maximizing the present value of economic rent from the timber.

The Pacific Northwest study compared future timber prices in the region for an assumed continuation of current federal forest policies with those for the assumption that public timber would be put on the market in response to market forces. The conclusion was that total timber harvests in western Oregon and western Washington would be about 18 percent higher during the 1980s and average timber prices about 80 percent lower than those that would occur under current federal timber-harvest policies. Over the following three decades, total timber harvests would continue to be from 6 to 20 percent above and timber prices from 60 to 70 percent below those under current policies. The volume of timber harvested from public lands, both federal and state, under this approach would be 2 to 2½ times that which could otherwise be expected during the 1980s and 1990s.

Since the above forecasts were made, conditions in the timber industry have changed substantially because of high interest rates and a depressed housing market. Forecasts based on current expectations for lumber and plywood markets over the next two decades indicate lower timber prices than those in the above forecast based on current federal timber policies.

The basic relationship between the two forecasts would, nevertheless, be the same.

It is important to understand that the lower prices brought about by more federal timber being offered for sale would apply to private, as well as public, timber if federal forestlands were managed to satisfy the same economic criteria as on private lands. The current value of privately owned timber inventories in the Northwest has been increased by public timber-harvest policies. This value would be severely eroded by any policy shift that would put substantially more of the present federal timber inventory on the market. That is, federal timber-harvest policies in the past have acted to increase significantly the income and wealth of private timber owners in the region. At the same time, these policies have restricted to a degree the ability of timber processors in this region to compete with those in the South and have limited employment in the region's timber-processing industry.

The overall economic and social impacts of current federal timber-harvest policies on a region such as the Pacific Northwest are difficult to weigh. Some parties in the region gain from such policies, while others lose. Federal land policies that affect receipts from federal land resources have created a complex set of real and expected benefits that are unequally spread among those in the public land states. Any shift in these policies would change the distribution of these benefits within the public land states and between them and the rest of the country.

Limitations on Administration of the Federal Lands

Federal land policies have long constrained the actions of the federal government as owner of the national forests and public lands. There is no doubt that Congress, under the "Property" clause of the Constitution, can do largely as it pleases with the federal lands. In fact, it has established, or has permitted the establishment of, a wide range of "restrictive arrangements" that limit what can be done with the federal lands. Many of them grew out of the customs and practices that have long been associated with the ways in which the lands are used.

Examples of these restrictive arrangements can be found for practically every use of the federal lands. Prospectors for hardrock minerals are free to search for minerals on the federal lands, which gives them a priority over other users. Grazing leases that ride with control over private property have become practically permanently assigned to long-standing permittees. The lands are available for outdoor recreation, hunting, and fishing without charge. The federal government has acceded to state laws for management of wildlife and water on the federal lands. It has even agreed, according to a recent opinion of the U.S. Attorney General, that it would henceforth subject itself on reserved federal lands to state laws for assigning rights to water, a reversal of earlier policies.

Most of these arrangements have favored those with specific interests in the federal lands, including the general population of federal land regions. This has been part of the price of having the lands held in federal ownership. For example, states and local governments have accepted a share of the revenues from the reserved federal lands in lieu of tax payments as one of the conditions of continued federal ownership.[25] In the absence of such arrangements over the past seventy-five years, the extent of the federal lands today would undoubtedly be much less. Two points deserve comment in this connection.

First, management of the federal lands is circumscribed by these restrictive arrangements. Changing a use when there has been a long-standing commitment of lands to particular uses or users is not easy. Some observers conclude that federal managers lack motivation to improve management, and others conclude that these same managers respond too readily to local interests. The reality of federal land management is far more complex than either conclusion suggests. Grazing permits cannot be changed readily to meet changes in range conditions. Access for hunting and fishing cannot be eliminated even if this would make timber management easier in some areas. Although the miner's actions may conflict with recreation, the rights of the miner to go on the federal lands cannot be abridged, except through complete withdrawals to mineral entry. The list of possible restrictions on management is a long one, and the restrictions are often subtle.

Second, any proposed changes in the status of the federal lands that would affect existing arrangements face resistance from interests that stand to lose. Since most of these arrangements have grown up to favor local interests in the western public land states, much of the resistance to change will come from this region. A classic example of this is the proposed, but never consummated, exchange of public domain timber lands in western Oregon in the early 1960s. The exchange would have put several tracts of public domain timberland in private ownership. It was part of a package to acquire national park lands at Point Reyes, California. On the surface, the concept of putting these lands in private ownership seemed sure to be supported by the timber industry. In the end, however, the proposal was fought, and defeated, by timber interests in Oregon because it would have disrupted their access to public timber.[26]

Cheaper on Fed Lands

Three specific changes in the extent of the restrictive arrangements affecting federal land management have occurred since 1970, the year that the Public Land Law Review Commission completed its broad review of federal land policies. This was also the year that the National Environmental Policy Act (NEPA) was signed into law and the federal land agencies began to grapple with its requirement for Environmental Impact Statements.

The first change was the initiation of full-scale, land use planning with the passage in 1976 of the Amendments to the Renewable Resources Planning Act of 1974 (RPA) and the Federal Lands Policy and Management Act (FLPMA). The second is the changes that have been made in payments to local governments that reflect the tax immunity of federal lands. The third is the increase in environmental protection requirements on the federal lands. The impact of each of these changes on management and tenure arrangements is substantial.

Land use planning

Planning serves at least two purposes for a public bureaucracy. It explains to users or other interested parties what the organization intends to do and it provides justification for programs. Despite the fact that their primary responsibility is land management, neither the Forest Service nor the BLM prior to 1970 prepared overall land use plans. A 1969 study for the PLLRC noted that typical land use planning by local governments is "markedly different in style and content" than that used by the federal land-mangement agencies. The study went on to note that land use planning outside of the federal government usually emphasizes "the preparation, public presentation and official approval" of comprehensive plans.[27]

This study was, in a sense, prophetic. Public presentation and official approval were, during the next decade, to become key elements in comprehensive land use planning for the federal lands. They also were to become two key factors in the disputes that have raged over planning for the federal lands.

The PLLRC placed considerable weight on land use planning in its 1970 report. It emphasized such planning as a way of accomplishing one of its major goals, to wit, forcing the land-management agencies to be more responsive to congressional direction.[28] Even in its own report, however, the commission did not really succeed in moving forward toward this goal. When it came to identifying the terms in which congressional direction would be defined, the commission waffled. It recommended that a land use planning process be established, but it was unable to say how priorities should be set and conflicts resolved.[29]

The commission's interest in land use planning was coincident with a growing interest in Congress in requiring land use planning by the states. One idea that had some appeal was that planning should not be required for private lands unless there was planning for federal lands. But the passage of NEPA really forced the issue.

The requirement of NEPA that Environmental Impact Statements be prepared for major federal actions forced the land-management agencies to prepare plans. In the absence of other plans, the Environmental Impact Statements themselves have served as the basic planning documents. NEPA also required that the agencies define "major actions" that affect environmental quality. The land-management agencies saw that land use plans were a vehicle that could be used to assemble many decisions in one place under the umbrella of one Environmental Impact Statement. Thus, by 1976, both the Forest Service and the BLM were well along in establishing land use planning procedures.

Congress faced much the same problem as the PLLRC when it came to passing RPA and FLPMA. That is, specifying acceptable goals for management of the national forests and public lands to be set in law proved to be more difficult in practice than in discussion. The goals of the RPA have been interpreted to be consistent with an overall mandate of "economic efficiency."[30] But if one judges by the results, it is not clear that RPA is being interpreted in Congress or by the agencies as mandating economic efficiency, at least where this interferes with other goals.

The Forest Service's planning process for the national forests has evolved in an expected direction, that of "rational" planning, wherein uses and land capabilities are measured, alternative programs defined, and a choice made among the alternatives. The BLM appears to be going in the same direction. In a more or less logical extension of rational planning, the Forest Service's complex FORPLAN computer program is being used on the national forests to make tentative land use allocations, which are then adjusted on the basis of other information, including that obtained through public involvement.

The evolution of the land use planning process for the federal lands has grown out of the desires of a wide range of interests. The PLLRC saw planning as a way to get greater responsiveness to congressional direction. Supporters of NEPA saw it as a way to show more consideration for environmental factors. The progressive conservationists saw it as "good government." And many of the interest groups saw it as providing another entry point to influence agency decisions. The common element in all of these viewpoints is that the documentation of decisions will result in better decisions.

Although the present planning process seems to have developed in response to a broad base of pressure for just such a process, it is not accepted enthusiastically in all quarters. Behan argues that the RPA mandate for, in effect, "perfect plans" will inevitably lead to paralysis of the land-management agencies through court actions.[31] Fairfax argues that the process takes decisions on land use away from experienced land managers and gives them to "lawyers, computers, economists, and politically active special interest groups" who, together, lack "on the ground" common sense.[32] Still others, while continuing to support the planning process, are frustrated because decisions are made slowly or they do not like the results.[33]

At least to some degree, the formal decision models, such as FOR-PLAN, that are used in national forest planning attempt to rationalize national and local interests in effect by asserting that they are identical. The models assign equal weights to national and local interests by subsuming them under criteria such as maximizing net benefits. The planning models might also be viewed as the latest in a long series of efforts by the national offices of the land-management agencies to maintain control over decisions in the field, an effort that will inevitably be countered by subsequent actions to put planners back on the ground.

Land use planning for the federal lands is still in a trial period. Nevertheless, the plans now being prepared will be seen as limiting what can be done with the lands and resources to actions that are consistent with the plans. In a sense, the plans are a contract with the public and with users that the agencies will take certain actions and not take others. Once they have been adopted, they will inevitably limit future changes in the direction of federal land programs.

Payments to State and Local Governments

Since 1906, payments have been made to states and local governments by the federal government as a replacement for taxes on federal lands.[34] The interest in these payments of local governments with large areas of federal lands was repaid during the 1970s with a substantial increase in the level of payments.

The primary way of making payments to state and local governments is to assign them a share of the revenues from the sale of land and resources. Thus, the level of payments varies with revenues, which go up and down from year to year and place to place, depending on market and resource conditions. Receipts from the sale of timber and from mineral leasing account for most revenues on national forests and public lands. Therefore, those states with the most timber and leasable minerals receive the highest payments. The other basis for making payments is "payments in lieu of taxes," which are intended to approximate taxes that would be paid on the property if it were in private ownership. Prior to 1976, payments in lieu of taxes were used mainly for federal lands that had been taxed at one time.

In 1970 the PLLRC recognized the burden imposed on state and local governments by the tax immunity of federal lands. It recommended that reliance be placed on payments in lieu of taxes, rather than revenue sharing, as the basic means for addressing this problem. The revenue-sharing arrangements at that time, in the view of the commission, failed to "meet a standard of equity and fair treatment either to state and local governments or to the federal taxpayers."[35] The revenue-sharing programs undercompensated in some cases and overcompensated in others. The standard that the commission sought to apply was that payments should relate to burdens imposed by federal ownership, and the major burden was the lack of tax payments. Thus, an equitable system of payments should compensate directly for the loss of taxes.

Since 1970, the level of payments to state and local governments has increased for four reasons. First, shared revenues have increased because prices for timber and energy minerals, sales of which are the basis for most of shared revenues, have increased more rapidly than prices in general. Second, Congress, in 1976, increased the percentage of revenues received under the Mineral Leasing Act that goes directly to the states from 37.5 percent to 50 percent. In that year, Congress also changed the basis for collecting revenues from coal leasing, and this increased these revenues sharply. Third, Congress decided in 1976 that the 25 percent share of national forest timber receipts that goes to local government should be based on gross revenues, including payments for regeneration of logged areas and the cost of purchaser-constructed roads.

Finally, in 1976, Congress provided that the states could, at their option, receive a flat payment in lieu of taxes instead of shared revenues from the federal lands. This is not a significant matter where revenues from timber, oil and gas, and coal are high, because the states choose revenue sharing instead of payments in lieu of taxes. It is, however, significant in areas used mainly for grazing and recreation or those with low-value timber.

In constant dollars, payments to the states in the eleven public land states from shared revenues for timber, oil and gas, and coal, and from payments in lieu of taxes increased 2.4 times from 1970 to 1980 (table

4-4). The increase for timber, where there was no change in the formula for revenue sharing, was 1.6 times. The increase for oil and gas and coal was 3.2 times. In current dollars, the total increase was from $146 million in 1970 to $732 million in 1980. Based on agency estimates of future receipts, total payments were estimated to increase to nearly $1.5 billion in 1985.[36]

The imbalance in payments among states, noted by the Public Land Law Review Commission in 1970, continues today. In 1970, 60 percent of the total payments from federal timber revenues in the eleven public land states went to the state of Oregon. In 1980 this had increased to 64 percent. Two states—New Mexico and Wyoming—get the bulk of payments from federal oil and gas revenues. Their share among the eleven public land states increased from 71 percent in 1970 to 74 percent in 1980, although there was some shift between the two states.

The payments-in-lieu-of-taxes provision passed by Congress in 1976 has done little to rectify the imbalance in federal payments among the public land states. Nevada, which has the highest proportion of its total area in federal ownership of any state, received $8.3 million in shared revenues from Forest Service and BLM lands in 1980, only 1.3 percent of the total for the eleven public land states. In addition, it received

TABLE 4-4. PAYMENTS TO ELEVEN WESTERN STATES
(millions of $)

Type of payment	1970	1975	1980
Revenue-sharing programs			
Timber[a]	100.0	154.6	345.8
	(114.0)	(121.3)	(187.5)
Oil and gas	41.3	79.1	278.6
	(47.1)	(62.0)	(151.1)
Coal	1.0	1.5	10.8
	(1.1)	(1.2)	(5.9)
Payments in lieu of taxes	—	—	76.4
			(41.4)
Total	142.3	235.2	711.6
	(162.3)	(184.5)	(385.9)

Note: Federal payments to states based on national forest and lands administered by the Department of the Interior. The eleven western states in this table are Arizona, California, Colorado, Idaho, Montana, Nevada, New Mexico, Oregon, Utah, Washington, and Wyoming. (Numbers in parentheses indicate millions of 1972$.)

Source: Office of Policy Analysis, "Past and Projected States Revenues from Energy and Other Natural Resources in Thirteen Western States" (Washington, D.C., U.S. Department of the Interior, 1981.)

[a]Current dollars are adjusted to 1972$ with the implicit GNP price deflator for government purchases of goods and services.

$5.3 million in payments in lieu of taxes, and this was only 6.9 percent of this total.

These figures strongly suggest that present revenue-sharing arrangements continue to overcompensate some states and undercompensate others, as the PLLRC noted in 1970. Tax immunity of federal lands is still an issue that deserves attention in terms of equity to the public land states and to the federal taxpayer.

The existing means of tempering the impacts of tax immunity raise barriers to changes in tenure arrangements and management objectives on federal lands. In practically every case where changes in land use or resource management have been proposed, the effect on payments to the states has been an issue. As long as payments are tied to revenues from the sale of federal resources, resource management on federal lands will be responsive to the interests of state and local governments. More often than not, they will be resistant to change.

Environmental restrictions

Since 1970, new environmental protection legislation has substantially increased the constraints on use of the federal lands. NEPA requirements for Environmental Impact Statements and consideration of environmental impacts in federal decisions are perhaps the most far-reaching by forcing a more public commitment by the land-management agencies to weigh environmental factors in their decisions. Requirements in legislation to protect endangered species, water quality, and air quality are not specific to the federal lands, but the public nature of decisions on these lands means that the land managers are more likely to be responsive than if the lands were not in federal ownership.

There also are some new environmental protection requirements that are specific to the federal lands. For example, while the rights of prospectors to go on the federal lands remain inviolate, regulations issued in 1974 for the national forests, and in 1980 for the public lands, require that proposed activities be described in operating plans if they would affect surface resources. These regulations are a commitment that the land-management agencies will take reasonable steps to limit impacts of mining on surface resources.

As with land use planning and increased payments to local governments, the new environmental requirements have added to the list of restrictive arrangements that constrain what can be done with the federal lands. There is, however, a possibly significant difference. The various restrictions—environmental and otherwise—and the benefits they provide apply to the federal lands as long as they remain in federal ownership. But if the lands were to pass out of federal ownership, most of the restrictions would presumably cease to be effective. This would not necessarily be the case with the environmental restrictions.

For one thing, Congress in the past has often attached restrictions to titles when it transferred land into private or other ownership. For example, acreage limitations have been used to favor small farmers in the disposal of public lands in reclamation projects, and transfers of land to states for recreation have been made contingent on their remaining in that use. Sustained-yield rules for timber harvesting have been attached to timberland that has been transferred to provide ownership. One might ask, Would environmental restrictions that apply more rigorously to federal lands transfer with ownership if such lands go into private ownership?

The Public Land Law Review Commission recommended that protective covenants be used in disposals of federal lands to protect the environment of lands that remain in federal ownership.[37] The particular case that concerned the PLLRC was land surrounding national parks, where unsuitable development could lower park values. The idea of extending parallel restrictions for environmental protection to lands on disposal into private ownership would certainly find supporters among the environmental interest groups.

In sum, restrictive arrangements on the federal lands have been adopted to protect or enhance someone's interests. Wholesale disruption of these constraints on management is unlikely. It is even possible that those that may find the fewest supporters among commodity interest groups, namely the environmental protection restrictions, would be the most likely to transfer with title if these lands were to go out of federal ownership.

Conclusions

Although vast in area, the federal lands are a limited resource. Overriding special use designations—national parks and monuments, wildlife refuges, wilderness areas, wild and scenic river areas, national trails—limit the area that is available for so-called multiple uses. Existing and potential resources on the remaining federal lands, the multiple-use national forests and public lands, generally provide only a small part of the nation's supply of goods such as minerals and forage for livestock and of services such as recreation, hunting, and fishing. They provide a larger share of the nation's timber and opportunities for dispersed recreation and downhill skiing.

The federal lands are also limited by the restrictive arrangements that have grown up around their use. These include arrangements that assign use of particular resources to specific parties, the public commitment of the agencies to specific courses of action through land use planning, and payments to state and local governments based on levels of resource use. Since 1970, environmental protection restrictions have increased in kind and impact. The effectiveness of these various restrictions in circumscribing decisions on the federal lands is substantial.

The important issue of federal lands policy is the degree to which the arrangements with users, local governments, and other special interests limit the contributions of these lands to economic and social objectives. Can use of the multiple-use lands be rationalized in the face of the existing constraints on their use? To some, the outlook is gloomy: land use plans will fix commitments for years in the future; or payments to states will make it impossible to achieve any changes in use that lower revenues. Some see the solution as disposal of the federal lands.

The PLLRC looked at the disposal issue some dozen years ago and concluded that retention of the federal lands would generally be in the best interests of the country. At the same time, it recommended an integrated revision of the then-current laws. The idea was that the package of revisions would be sufficiently attractive in its entirety so that the barriers imposed by a host of restrictive arrangements could be overcome.

This hope soon proved to be wrong in the debates following the PLLRC report over "dominant use," which the commission saw as a kind of integrating management concept.[38] In the end, many of the PLLRC recommendations were adopted, either by Congress or by the executive branch. But they were adopted in a piecemeal fashion, which to a degree has replaced one set of conflicting policies with another.[39] The many special arrangements that had accumulated around the use of federal lands barred the way to a wholesale change in policies.

There is little reason to think that the proponents for disposal of the federal lands will be any more successful. To a large degree, we are caught in our own history, the evidence of which is in the many special arrangements that have accumulated around the use of the federal lands. Each of them represents some form of payment for the present body of federal land policies. We cannot wish them away. They will continue to limit what can be done with the federal lands.

Notes

1. Marion Clawson, *The Federal Lands Revisited* (Washington, D.C., Resources for the Future, 1983); Robert H. Nelson, "1978 Revenues and Costs of Public Land Management by the Interior Department in Thirteen Western States" (Washington, D.C., Office of Policy Analysis, U.S. Department of the Interior, 1980); and *Federal Coal Policy: A Crisis in Public Resource Management* (Durham, N.C., Duke University Press, In Press).

2. U.S. Bureau of Land Management, *Public Land Statistics* (Washington, D.C., Government Printing Office, 1980) pp. 44–45.

3. Clawson, *The Federal Lands Revisited*.

4. U.S. Public Land Law Review Commission, *One-Third of the Nation's Land* (Washington, D.C., Government Printing Office, 1970).

5. John D. Black, in James P. Cavin, ed., *Economics for Agriculture* (Cambridge, Mass., Harvard University Press, 1959) pp. 137*ff*.

6. U.S. Forest Service, "An Analysis of the Timber Situation in the United States: 1952–2030" Draft (Washington, D.C., U.S. Forest Service, 1980).

7. William F. Hyde, *Timber Supply, Land Allocation, and Economic Efficiency* (Baltimore, Md., Johns Hopkins University Press for Resources for the Future, 1980) chap. 5; and "National Forest Logs Red Ink for Treasury," *Wharton Magazine* vol. 6, no. 1 (1981).

8. U.S. Bureau of Land Management, *Public Land Statistics*, p. 165; and U.S. Forest Service, "Report of the Forest Service: Fiscal Year 1980 Highlights" (Washington, D.C., 1981) p. 34.

9. Paul J. Culhane, *Public Lands Politics: Interest Group Influence on the Forest Service and the Bureau of Land Management* (Baltimore, Md., Johns Hopkins University Press for Resources for the Future, 1981).

10. Clawson, *The Federal Lands Revisited*.

11. Ibid.

12. PLLRC, *One-Third*, p. 105.

13. Culhane, *Public Lands Politics*.

14. U.S. Forest Service, *An Assessment of the Forest and Range Land Situation in the United States* (Washington, D.C., Government Printing Office, 1981) p. 162.

15. PLLRC, *One-Third*, pp. 136–138.

16. U.S. Office of Technology Assessment, *Management of Fuel and Nonfuel Minerals in Federal Land* (Washington, D.C., Government Printing Office, 1979).

17. J. H. Powel and Gary K. Loth, *An Economic Analysis of Nontimber Uses of Forest Land in the Pacific Northwest*, PB 109182, prepared for Forest Policy Project, Washington State University (Springfield, Va., National Technical Information Service, 1981) pp. 63–66.

18. Roger D. Fight, K. Norman Johnson, Kent P. Connaughton, and Robert W. Sassaman, "Roadless Area Intensive Management Tradeoffs on Western National Forests" (Washington, D.C., U.S. Forest Service, 1978).

19. William E. Bruner and Perry R. Hagenstein, *Alternative Forest Policies for the Pacific Northwest*, PB 82109125, prepared for Forest Policy Project, Washington State University (Springfield, Va., National Technical Information Service, 1981).

20. Harold K. Steen, *The U.S. Forest Service: A History* (Seattle, University of Washington Press, 1976) pp. 122–132.

21. Clawson, *The Federal Lands Revisited*.

22. Interdepartmental Grazing Fee Committee, "Review of Federal Land Administration for Livestock Grazing" (Washington, D.C., U.S. Departments of Agriculture, De-

fense and Interior, 1967). Referenced in University of Idaho and Pacific Consultants, Inc., *The Forage Resource*, PB 189 250, prepared for U.S. Public Land Law Commission, Washington, D.C., (Springfield, Va., National Technical Information Service, 1970) p. III-40; U.S. Forest Service, "Grazing Fees on National Forest Range" (Washington, D.C., 1968); and Thomas R. Waggener with Joseph M. McDonald and Thomas C. Lee, *User Fees and Charges for Public Lands and Resources*, prepared for U.S. Public Land Law Review Commission, PB 195 846 (Springfield, Va., National Technical Information Service, 1970) p. 324.

23. PLLRC, *One-Third*.

24. Bruner and Hagenstein, *Alternative Forest Policies*, p. 6-64.

25. Paul W. Gates, *History of Public Land Law Development* (Washington, D.C., Government Printing Office, 1968) pp. 28–30.

26. Elmo Richardson, *BLM's Billion Dollar Checkerboard: Managing the O&C Lands*, published for the Forest History Society (Washington, D.C., Government Printing Office, 1980) pp. 154–161.

27. Herman D. Ruth & Associates, *Regional and Local Land Use Planning*, PB 189410, prepared for U.S. Public Land Law Review Commission, Washington, D.C., (Springfield, Va., National Technical Information Service, 1969).

28. PLLRC, *One-Third*, pp. 41–42.

29. Ibid., pp. 45–48; and Perry R. Hagenstein, "One-Third of the Nation's Land— Evolution of a Policy Recommendation," *Natural Resources Journal* vol. 12, no. 1, (1972) pp. 56–75.

30. Ibid., pp. 45–48; and PLLRC, *One-Third of the Nation's Land*.

31. R. W. Behan, "RPA/NFMA—A Time to Punt," *Journal of Forestry* vol. 79, no. 12 (1981) pp. 802–805.

32. Sally K. Fairfax, "RPA and the Forest Service," in William E. Shands, ed., *A Citizen's Guide to the Forest and Rangelands Renewable Resources Planning Act* (Washington, D.C., U.S. Forest Service/Conservation Foundation, 1981).

33. Carl A. Newport, "A Foregone Conclusion About Timber Increases," *Journal of Forestry* vol. 80, no. 3 (1982) p. 160.

34. Gates, *History*, p. 29.

35. PLLRC, *One-Third*, p. 237.

36. Office of Policy Analysis, "Past and Projected Revenues from Energy and Other Natural Resources in Thirteen Western States" (Washington, D.C., U.S. Department of the Interior, 1981).

37. PLLRC, *One-Third*, p. 82.

38. Hagenstein, "One-Third of the Nation's Land."

39. Jerome Muys, "The Public Land Law Review Commission's Impact on the Federal Land Policy and Management Act of 1976," *Arizona Law Review* vol. 21, no. 2, (1979) pp. 301–309.

5

Uses and Limits of the Federal Lands Today

Who Cares and How Should Current Law Work?

D. MICHAEL HARVEY

> The astounding value of the public domain lies in uses that have been
> overlooked, for the most part . . . as deterrents to floods, uses for natural
> storage for irrigated lands, uses for game and recreation, uses for forage
> and range, uses as forest areas, and finally uses as gigantic regulators of
> climatic conditions . . . the value of these lands is such that the welfare
> and prosperity of the people of a mighty empire all depend upon the fidelity
> with which the lands we call the public domain are administered.
> —Rep. Burton French (R–Idaho) 1934

The federal lands have always provided the arena in which we Americans
have struggled to fulfill our dreams. Even today dreams of wealth,
adventure, and escape are still being acted out on these far-flung lands.
These lands and the dreams—fulfilled and unfulfilled—which they foster are a part of our national heritage. What we do with them tells a
great deal about what we are, what we care for, and what is to become
of us as a nation.

The purpose of this workshop is to consider tenure arrangements on
the federal resource lands administered by the Forest Service and the
Bureau of Land Management (BLM). As Perry Hagenstein has so ably
pointed out, while these lands are vast, they are limited in the contributions they are likely to make to society. I agree with him that there
are two types of limits—physical and institutional.

Using Hagenstein's description of the current uses and limits of the federal lands as a starting point, I want to focus on two critical questions that I believe should be considered before possible changes to existing tenure arrangements.

Two Critical Questions

The first question is, Who cares, or what different groups or institutions have an interest in these lands? On this question my analysis applies to both national forests and public lands administered by the BLM.

The second question is, How effective are current laws governing management of these lands, and would different tenure arrangements be better for the groups or institutions affected? On this question my analysis will focus on the public lands administered by the BLM. Although most of my comments would for the most part be also applicable to the national forests, I believe that the principal target of the current debate over land-tenure arrangements, particularly transfers out of federal ownership, are the public lands and, therefore, it is appropriate to devote the bulk of our attention to them. The national forests and the Forest Service have enough public recognition and appreciation to guarantee careful evaluation of any proposed change in tenure arrangements. The public lands and the BLM do not.

During the 1970s Congress established significant new policies for management of the federal lands. These include the Forest and Rangeland Renewable Resources Planning Act of 1974, the National Forest Management Act of 1976, and the Federal Land Policy and Management Act of 1976. These policies are just now beginning to be fully implemented.

At the same time, some people question the very concept of federal land ownership. Beginning in the late 1970s, the Sagebrush Rebellion called for the federal lands to be given to the states. After that notion failed to take hold, many of the rebels now call for privatization of the federal lands. Apparently, this means transfer of title to all the lands to private individuals, although the size, timing, and terms of these transfers are not specified. Nor is it clear who will be eligible to purchase, or what will happen to existing users.

This uncertainty is compounded by the fact that proponents of privatization, including administration representatives, use the term in two ways. One usage refers to the accelerated sales program—asset management—apparently primarily designed to raise revenue "to retire the national debt." The other usage reflects a philosophical view that private ownership of federal lands would result in net national economic benefits because private management is more efficient than public management.

Many people, including Marion Clawson, have suggested other, more imaginative ways to increase private control of currently federal lands.

The Conservative Approach

In this volume we consider possible alternative future policies for the federal lands—some of which would involve radical change. My views on all such alternatives are conservative.

Why conservative? First, I prefer evolution to revolution. Second, I believe in private use of public resources subject to market forces, but also believe that public ownership protects from generation to generation vital, long-term resource values which the market place does not recognize. Third, I believe that land in general, and federal land in particular, has values that are not, and indeed cannot be, recognized in real estate sales. Federal lands have values that cannot be measured in dollars. Fourth, I believe in conservation of natural resources, which I define as "wise use." I believe that wise use of our natural resources must attempt to balance all the different, and sometimes conflicting, uses with the ultimate resolution being that expressed by Gifford Pinchot almost eighty years ago—"The question will always be decided from the standpoint of the greatest good of the greatest number in the long run."

Pinchot's famous maxim was rephrased by Congress in the Public Land Law Review Commission Act in 1964, which called for public land management to achieve "the maximum benefit for the general public." Of course, it is not so easy to determine what that maximum benefit is in any given situation.

Who Cares?

I find the commission's identification of six "publics" that together make up the general public to be useful tools in analyzing public land policy issues. These are the national public; the regional public; the federal government as sovereign; the federal government as proprietor; state and local governments; and the users of the public lands. What are the interests of those "publics"?

The national public

Although the federal lands are not distributed proportionally through-out the nation, they and their resources belong to all the people of the United States.

The national public has an interest in reducing the burden on taxpayers generally, either by maximizing the net revenue from the federal lands, or by assuring more efficient management, or both. The national public also has an interest that consumer goods and services derived from the federal lands will be made available at the lowest possible price con-sistent with good conservation and management practices. In other words, the federal lands should be managed efficiently, and all development costs should be internalized.

The national public is concerned that the federal lands should con-tribute to the maintenance of a quality environment and that the federal lands should be managed to enhance human and social values. While the interests of the national public are not associated with any particular kind of use of the federal lands, the national public is concerned that people who do use these lands be treated equally.

The regional public

Those who live and work on or near the federal lands have separate, identifiable, and special concerns that go beyond their interest as mem-bers of the national public. They have a strong desire that these lands contribute meaningfully to the quality of the environment in which they live.

Taxes on private property ownership are a major source of revenue in federal land states, particularly at the local level. It is in the regional public interest to have the federal government, as a landowner, pay its fair share of the costs of adequate local and state governmental services.

Federal lands and their resources are an important part of the eco-nomic base in at least twenty-two states. There clearly is a regional public interest in laws and policies which permit federal lands and their resources to contribute to regional growth, development, and employ-ment.

The federal government as sovereign

As a matter of constitutional law, there is no legal significance in the different roles of the federal government as sovereign and as proprietor, but it is useful to separate these two institutional interests in federal land. By so doing, we may distinguish those interests which relate to governmental functions from those which are similar to the interests of any other landowner.

If the federal government is to achieve its broad constitutional responsibilities toward the national community, federal land laws and policies should complement and implement other nationwide programs and policies. Federal lands must be viewed as one of the tools that the federal government has available in pursuing its sovereign objectives.

The federal sovereign interest lies in the efficient economic and non-economic utilization of all the resources of our nation and the avoidance of diversion of labor and capital to less-productive enterprises. Consequently, from the sovereign point of view, laws and policies should not permit federal lands and resources to be used in unfair competition with resources from other sources. Withholding of federal land resources from development may in different circumstances either further or thwart the sovereign interest.

The national interest requires users of federal lands and resources to contribute their fair share of federal revenues. This principle precludes tax or pricing policies which unduly favor the users of federal lands. There is a sovereign interest in assuring access on equal terms to all potential users of the goods and services from those lands. The avoidance of monopoly and special privilege is the basic policy of many federal laws, including, for example, the antitrust laws.

There is also a sovereign interest in the maintenance of quality environmental conditions on federal lands at least equal to those standards legislated for the nation generally. It would be unfair to enforce on the private sector standards higher than those established for federal lands by the very government charged with their enforcement.

The federal government as proprietor

In its role as proprietor, the federal government has much the same interest as other landowners, and wants at least the same degree of freedom to manage and use its resources.

As a proprietor, the federal government wants to maximize the net economic return from sales of land and resources.

The government, in the role of proprietor, has an interest in assuring the availability of sufficient funds to finance programs at a level that will result in a new monetary gain. It is also interested in the furtherance of research to achieve better use of the land.

The federal proprietor, in addition, has an interest in controlling users of the land in order to maintain the resource base and minimize damage or adverse environmental impacts.

State and local governments

In the absence of conflicting federal legislation, state and local governments have constitutional jurisdiction over federal lands for many purposes except where exclusive federal jurisdiction has been ceded over specific areas. State and local governments have an interest in obtaining an equitable share of their governmental costs from the federal government as a proprietor of federal lands.

State and local governments that will be affected by land use decisions expect that they will be consulted and have a voice in the federal decision-making process. They expect the United States in that way to give consideration to relevant state and local programs and also to consider the impact of federal land actions on state and local governments.

Because they use federal lands for public purposes, these units of government expect a preference over competing potential users, and to purchase or lease federal land at less than market value.

Users of federal lands and resources

Those who use the federal lands as a basis for economic enterprise and those who use the federal lands for noncommercial purposes, such as personal recreation, have an identifiable interest in the public lands. This is not necessarily a short-term interest, since all users are concerned that federal land policies provide an opportunity for the satisfaction of future requirements as well as present needs.

While users as a group have a common interest in the federal lands, different classes of users, and, indeed, individual users within classes, often must compete for the opportunity to use the federal lands. Many of the controversies over federal land policy involve such conflicts, and they should be so recognized.

Users want equal opportunity for access to federal lands and resources in which they are interested, and equal treatment in their relations with the federal government and with other users. They are interested in having a voice in decision making from the time that plans are made for general use through the chain of events that may involve decisions affecting their particular uses. In this latter connection, of course, all users desire prompt and fair consideration of disputes with federal land administrators.

Users are interested in having the terms and conditions under which use will take place specifically stated in advance. Although such need is not always recognized by those who use the federal lands for noneconomic purposes, it is significant and should be taken into consideration.

Users also have a justifiable interest in seeking pricing and other conditions competitive with the use of other lands, together with security of investment, usually through assured tenure of use. As a corollary, they expect to be compensated if their use is disrupted or interfered with before the expiration of the term of the lease or use permit.

Current Public Land Management

Background

With those considerations in mind, let me first review the public lands and the new policies, and then suggest a course for the future.

As a nation, Americans once owned nearly 2 billion acres of public lands, but in the course of our national expansion and development, they were sold or given by the federal government to the states and their counties and municipalities, to educational institutions, and to private citizens and industries. Other lands were set aside for national parks, forests, and wildlife refuges, and for military installations.

About 350 million acres remain—one-fifth of our nation's land area. A little more than half are in Alaska. In the "lower 48" the other 174 million acres are almost entirely in the eleven western states.

Although they are rarely recognized as such, many of the most pressing domestic policy issues the nation currently faces are also public land issues: oil and gas development, energy facility siting and transcontinental pipelines, coal development, synfuels development, nuclear waste disposal, MX missile deployment, timber production, the availability of water and land for urban development, and the erosion and disappearance of the nation's open public spaces.

At the same time, few areas of the country are so vast, so rich, and so poorly understood as the public lands. Few Americans understand that they own all this land, or know where it is, what it is worth, who is in charge, or why it is managed the way it is.

As we look for ways to meet our growing energy, food and fiber, timber, water, recreational, and other demands, national attention has turned to the resources of the public lands owned by all Americans and managed by the secretary of the interior through the Bureau of Land Management (BLM). The attention is long overdue. Not surprisingly, controversy and conflict over the use of these resources have followed. The reason is obvious. We recognize the public lands today as a vast storehouse of resources and values that we will depend upon to sustain us in the next century.

Resource values

The public lands were until recently called "the lands nobody wanted." But those days are over. The competition for the use of these lands is intense and growing and, given what is at stake, that is no surprise.

• Recent estimates by the BLM and the U.S. Geological Survey are that the public lands contain 25 percent of the nation's oil and gas, 50 percent of its coal, 80 percent of its oil shale, 50 percent of its uranium, and 50 percent of its geothermal energy. Management of these public resources is arguably the most important energy policy responsibility of the federal government.

• In addition, the public lands contain an enormous storehouse of the nation's nonenergy mineral deposits, including some of the world's principal sources of berylium, molybdenite, antimony, phosphate, and potash.

• Four percent of the nation's beef cattle and 38 percent of its sheep graze on the public lands at some point during their lives.

• About 1.2 billion board-feet of timber—enough to build 86,000 three-bedroom houses—come from the public lands in an average year.

• Nearly 200 million people visit literally hundreds of unique recreational areas on the public lands.

• The public lands are home to one out of every five of the big-game animals in the entire United States, including most of the caribou, brown and grizzly bears, and desert bighorn sheep, 50 percent of the moose, 65 percent of the mule deer, and 45 percent of the antelope. No single federal or state agency manages more wildlife habitat than the BLM.

Recent changes

These are our "national resource" lands. Because of their richness and diversity they provide a classic formula for conflict. Historically, these lands were used predominantly by western livestock operators and mining interests, with few rules or competing demands. Recently, however, awareness of the importance of these lands for a much wider variety of uses has increased substantially. Increasing scarcity of basic commodities; a rapid shift of the urbanized demographic center of the nation to the West and South; and a growing appreciation of environmental values have increased pressures for the use of these lands for such competing purposes as energy production, urban growth, recreation, wildlife management, and wilderness preservation—as well as for the traditional uses. These demands have broadened the range of people who feel they have a stake in, and something to say about, how the public lands are managed.

It is not so much that the rules of the public lands management game have changed as it is that there are many new players. There are new definitions of the public interest, new constituencies to be served, and new claims on the resources of the public lands.

Congress has been the leader in recognizing this evolution of demands, values, and attitudes. During the 1970s it enacted a number of laws that impacted public land management including requirements for environmental impact analyses; changes in federal oil, gas, and coal leasing; regulation of surface coal mining; protection of endangered species; designation of public lands in Alaska as parks or other conservation system units; and improvement of public rangelands.

Federal Land Policy and Management Act

The most important and sweeping of these laws is the Federal Land Policy and Management Act of 1976 (FLPMA). This act replaced over 2,500 individual laws and was the first definitive, comprehensive statement of public land management policy. It formally reversed the long-standing view that the public domain lands would eventually be disposed of; it mandated that they be retained in federal ownership for the benefit of the entire nation, unless it was in the public interest to dispose of a particular parcel.

FLPMA sets out criteria for sales and exchanges of public lands. Clearly, Congress anticipated that significant acreage would be transferred out of federal ownership. Estimates in the 1960s indicated that at least 5 million acres probably would meet these criteria.

FLPMA established that the public lands are to be managed under principles of multiple use and sustained yield. In so doing it expressly recognized that the many and varied resources of the public lands are important but not limitless. They require balanced use to realize their many potential benefits.

FLPMA did not specify which uses would be allowed on which areas. Rather it set out a planning process to be used for making land use decisions and general principles to be followed in those decisions.

The basic multiple-use–sustained-yield principle calls for

. . . Management of the public lands and their various resource values so that they are utilized in the combination that will best meet the needs of the American people . . . a combination of uses that take into account the long-term needs of future generations for renewable and nonrenewable resources, including, but not limited to, recreation, scenic, scientific and historical values . . . harmonious and coordinated management of the various resources without permanent impairment of the productivity of the land and the environment with consideration being given to the relative values of the resources and not necessarily to the combination of uses that will give the greatest economic return of the greatest unit output.

The basic principles to be followed include (1) decisions based on resource inventory and land use planning with broad public participation; (2) maximum protection of the environment, with the cost of preventing or minimizing damage paid by users; (3) receipt of fair market value for private use of public resources; and (4) close cooperation with state and local governments and local citizens and consistency with state and local land use plans.

During the six years that Congress was considering FLPMA prior to its passage and approval by President Ford in October 1976, these general principles and the land use planning process were accepted by both preservationists and developers.

FLPMA was proposed by western Republicans in the executive branch. The congressional leaders on the measure were all westerners. Any one senator could have stopped its approval on the last night of the Ninety-fourth Congress. No one did.

Implementation of FLPMA

However, implementation of the new law by the Carter administration started a fierce debate. This debate is frequently characterized as being over the "choice" between preservation and development. Some of the industry press statements might lead one to believe that Secretary Andrus, who had been governor of Idaho, had become a tool of "eastern environmental elitists." As a result, he "locked up" the public lands, preventing them from being expeditiously developed, and thereby frustrating industry efforts to (1) decrease the nation's dependence on foreign energy and mineral supplies, and (2) provide lumber for housing.

At the same time, environmental groups felt that development plans were being allowed to move ahead without proper attention to wilderness potential and natural values of the land. They cited the fact that more than 100 million acres of public land were already under oil and gas lease but 80 percent had not been explored. They cited the rapid expansion of production from federal coal leases with 25 billion tons under lease. They alleged continued overgrazing of domestic livestock and failure to control environmental damage from mineral development under the still-operative 1872 Mining Law.

These philosophical and political debates between "environmental protection" and "resource development" usually ignored the central issue: human survival does not depend on one or the other—it depends on both. FLPMA recognized that fact. The law does not mandate one or the other on any or all public lands. It recognized that both are necessary, and achievable.

More and more frequently these issues were also characterized as East versus West. The ultimate expression of this view is the so-called Sagebrush Rebellion. The goal of the Rebellion is federal legislation directing the transfer of the public lands to the states. The Rebellion had a rapid rise to national prominence, perhaps reaching its high point in late 1980, when then-candidate Ronald Reagan said, "Count me in as a rebel"; and Secretary of the Interior-designate James Watt said, "I am a Sagebrush Rebel."

Since both took office the Rebellion has waned considerably. There have been no hearings on the legislation, and former Secretary Watt believed his "good neighbor" policy with state and local governments made the transfer unnecessary. Any administration, particularly one interested in budget balancing, would be loath to give up the large revenues from these lands which far exceed budgeted management costs. Perhaps they have come to share the view, held by many, that the Rebellion is one last effort by livestock and mineral interests to keep effective control over the public lands—the "lands nobody wanted" in the face of new demands and new users, many of them from local and regional residents. I share John Leshy's view that the Rebellion will ultimately be viewed as "the last gasp of a passing era, an effort to turn back the clock to the days when competition among users of public lands was rare, when resources seemed inexhaustible, and when a consensus existed for exploitation."

Now some in the administration favor privatization. The president's budget assumes substantial revenue from sales of public lands. There has been talk of possible new legislation, but none has been forthcoming.

Making Management Effective

Where should we go from here? Congress wisely did not establish an either–or policy for public land management. It recognized the diversity of resources and the varied, sometimes competing, demands for their use. Congress anticipated that professional public land managers operating on-the-ground would apply the policies and the land use planning process spelled out at the national level in FLPMA to strike a balance among uses and users.

It has been said that "we do not inherit the land from our parents, we borrow it from our children." FLPMA requires that this balance also has to be struck among generations. The private marketplace, for all its virtues, is not equipped to deal with the various long-term, virtually nonquantifiable, values of the public lands, such as watershed protection and wilderness preservation.

How can this elusive balance be achieved? First, we must implement the land use planning and decision-making process established by FLPMA. We do not need new public land laws. We may very well need to streamline existing regulations and procedures based on experience. Existing law allows this. We do need to get on with making the hundreds of thousands of individual decisions that are required every year. Some of these decisions will be to sell or exchange public lands.

Decisions must be based on full information on resource capability and on the costs and benefits of alternative uses. There must be full representation of conflicting interests and viewpoints. This means not only interests like timber, mining, recreation, and wildlife, but also urban consumers, state and local government, local industries, and neighboring landowners. Once made, decisions should be carried out so that land users can safely make investment decisions. Most users are ready and willing to take steps, sometimes costly ones, to prevent or minimize damage to the land and the environment from their activities. They need to know what is required and believe that the requirements will be enforced reasonably, but promptly.

Second, the executive branch and Congress must make a commitment to the FLPMA processes. This means providing adequate manpower and funding so that professional land managers can do their job properly. It also requires delegation of decision-making authority to the lowest possible level in the field. Congress and political appointees in the executive branch must provide adequate guidance for exercise of that authority. They must also refrain from second-guessing decisions of field managers unless they are clearly inconsistent with established law, regulation, or policy.

A large portion of the public lands budget consists of items that any responsible private business would treat as capital investments, not annual operating costs. It is time that the federal government do so, too. A portion of the annual income from these lands should be set aside as a public land capital improvement fund.

Third, in an era of reduced federal budgets and a widespread desire to return power to the states, we should take full advantage of all the federal–state cooperative provisions of public land policy. These include consistency between federal management plans and programs of state and local land use planning; opportunities for state and local governments to acquire public lands needed for public purposes, frequently at little or no cost; and fifty-fifty sharing of federal mineral leasing revenues with the states, with priority for expenditure of the state share to aid local governments impacted by federal mineral development. In 1983 these payments are estimated to be close to $700 million.

Finally, and most important, we must reduce the polarization that has characterized much of the recent public land policy debate. Cecil Andrus called traditional public land users the "rape, ruin, and run boys." Former Secretary Watt referred to "toadstool worshippers" and "environmental extremists," who are always selfish, usually easterners, and possibly Communists. These and other colorful statements make good press, but they do not help to make good public land decisions.

We all must recognize that Congress did not side with any user group. It took the middle ground, not just out of political expediency but out of recognition of reality. Public land issues are never black and white. Neither are they simply development versus preservation, East versus West, urban versus rural, public versus private, or federal versus state. They are all these and more, wrapped up together in a complex web.

All interested parties must be willing to enter into dialogue, not diatribe. We need reasoned responses, not knee-jerk reactions; cooperation, not confrontation or litigation. All must be willing to work within the FLPMA process to make their arguments and resolve their differences. To ignore them or to propose radical changes in the existing principles or processes is to invite suspicion, stridency, and eventually, I think, lead to stalemate.

If the FLPMA process is short-circuited or starved before it has a chance, we will all be losers. Instead of cooperation that is necessary to make the process work, we shall see more of the polarization in which "protectors" and "users" try, both in Congress and the courts, to throw monkey wrenches into each other's machines. Any hope for the certainty needed to preserve land or to develop it wisely will disappear. The risks of paralysis or backlash will grow.

Let us use the policies and processes of FLPMA which are expressly designed to achieve soundly balanced public land programs. If we do, we can achieve the goal of Congress that the public lands be managed in a manner that "where appropriate, will preserve and protect certain public lands in their natural condition." At the same time we will "recognize the nation's need for domestic sources of minerals, food, timber, and fiber from the public lands" while also transferring out of federal ownership by sale or exchange significant acreage, so as to achieve the basic objective that these lands "are utilized in the combination that will best meet the present and future needs of the American people."

III
Retention, Disposal, and Public Interest

6

The Claim for Retention of the Public Lands

JOSEPH L. SAX

In the spring of 1982, a proposal put forward by the Reagan administration to sell a substantial fraction of the federally owned lands (some 35 million acres was the figure most commonly mentioned) attracted considerable attention. By midsummer, the issue had become a leading news item, reaching the first page of the *New York Times* and the cover of *Time* magazine. As summer waned, so did the land sale proposal. Its principal proponent in the White House, Steve Hanke, had gone back to his academic post after strong reservations to the proposal had been expressed by numerous spokesmen for western public land states; and journalists began to turn their attention to other, more newsworthy issues.

There are sufficient practical and political reasons to make it unlikely that any large-scale sale of federal lands is in the offing. But the issue is a perennial in American political discourse because some profound issues of public policy are raised by substantial public land ownership and management in a nation deeply committed to private proprietorship of major resources and industries. It is those issues to which I intend principally to address myself in the pages that follow. Before turning to that, however, I propose to make a few preliminary observations about some of the practical questions raised by current proposals for the disposition of public domain lands.

Despite some of the flamboyant journalistic accounts of the issue, there is nothing in any plan put forward by the Reagan administration

125

to suggest that the national parks might be put on the auction block or that the administration plans to dispose of designated wilderness areas. Whether there is some secret and nefarious agenda behind the public words is beyond my knowledge, but the fact is that, to date, discussion has been limited to unspectacular lands largely used for commodity purposes (such as grazing) held by the Bureau of Land Management (BLM) and the Forest Service. However obvious it may be to many people that the United States should own and control Yellowstone National Park or the Lincoln Memorial, it is far from self-evident why the United States should also own and control hundreds of millions of acres of rather ordinary land.

I do not assert that such lands should be sold, but I think the question is one that deserves serious consideration. It should be noted here that federal ownership is not based on any long tradition or constitutional principle. For many years it was everywhere assumed that the United States would dispose of its landholdings, and it did dispose of much of what it owned until the mid-1930s. As late as the 1890s it was still considered a serious constitutional question whether the United States could acquire land for the purpose of establishing a memorial to the Civil War.[1] But it was not until 1976 that the Congress, in the Federal Land Policy and Management Act (FLPMA), declared its general intention to retain in permanent federal ownership the remaining public lands, which then constituted about three-quarters of a billion acres.

In the light of this rather recent commitment to permanent public ownership, one might wonder why there is such clamor at the prospect of disposition. Alternatively, since the federal government has, in fact, owned so much of the West for so long, one might equally wonder at the sources of enthusiasm for disposition. I suggest that several practical reasons have engendered the current controversy.

The first is symbolic and strongly related to the identity of the proponents of disposition. FLPMA was enacted at a time of intense interest in environmental protection and contains (among many other things) strong provisions dealing with wilderness and the management of areas of critical environmental concern. The Reagan administration is widely perceived as being indifferent or even hostile to such interests. It is not surprising then that its sale proposals are taken as a signal of that indifference. While—as I shall explain more fully below—no necessary relationship exists between proprietorship and the imposition of environmental controls on land use, disposition would be an easy way to shuck off such responsibilities. There was little in former Interior Secretary Watt's behavior or rhetoric to allay such anxieties.

A related concern—and for the same reasons—is the fear of what is usually called a "giveaway." The administration has taken pains on numerous occasions to identify itself as a friend of private enterprise. It is hardly astonishing if there is some fear that land sales might be made at less than their full value causing what some fear would be a redistribution of wealth in favor of the rich, and not merely a change in the form of capital from land to cash. The fact that laws can be enacted to demand fair and full value for any sales does not fully allay such fears. Appraisal of land values is an inexact art, and its inadequacies are intensified in areas where sales are relatively infrequent or where the lands in question do not have exact counterparts. It is easier to appraise a conventional house in a suburban subdivision, for example, and more difficult to appraise Yosemite Valley. Most of the lands currently in question lie between these extremes.

It is not irrelevant to note also that the history of public land disposition in the United States is replete with instances of nefarious dealing. Nor is it irrelevant to ask what mechanisms might be imagined to ensure against a resurgence of such problems. One of the many ironies in the debate over the public domain is that those who favor sales are largely those who are most skeptical of the performance of federal bureaucrats as managers. How can they be confident that these same bureaucrats as land sellers would perform satisfactorily?

Concerns about the practical outcome of land sales have not been alleviated by the suggestion that the proceeds of such sales should be applied to the payment of the public debt. (It should be noted that whether it is desirable to diminish the public debt is an entirely separate question from whether the public domain should be sold. The debt could be reduced by diminishing defense spending, or selling aircraft carriers, or by reducing welfare payments. There is no necessary relationship between the desire to lessen a debt and the identity of the resources used to achieve that end.) Emphasis on the size of the public debt as a reason for selling public lands has probably stirred, rather than allayed, suspicions as to the motives of the administration. Since carrying charges on the debt come largely from tax revenues that are progressive, it is quite possible that the principal beneficiaries of debt reduction would be high-bracket taxpayers. It is possible, of course, that the bulk of current beneficiaries of the public lands are taxpayers in equally high brackets, but that is far from clear. I do not purport to know the facts, but the possibility exists that a land-sale–debt-payment project would result in a redistribution of wealth favoring the wealthy.

Claims that significant reductions of the public debt would result from land sales have caused confusion of another kind that has engendered suspicion of the administration's plan. If, as has been said, the government only intends to sell millions of acres of low-value grazing, forest, and desert lands, the proceeds are hardly likely to make much of a dent in the trillion-dollar national debt. The big money is in the high-value lands, those rich in mineral, timber, and recreational resources. If those are the lands in question, quite different public responses might be expected.

Another wholly different matter has restrained enthusiasm for the land-disposal project, and, in my opinion, it is likely to be decisive against any large-scale sales. It is simply the fact that the existing arrangements for use of the public domain reflect decades of political and economic bargaining under which many interest groups of any consequence have obtained benefits that suit them and to which they have adapted. The existing system, however imperfect it may appear from the perspective of economic efficiency or resource management, is, after all, a "mature" system that incorporates some highly developed expectations. Changes that unsettle long-held expectations are, simply by virtue of that fact, costly. This is not to assert that change should never be made; it is to say only that change imposes its own costs, and that it is unrealistic to expect people to respond to any ideas proposed for an established system as if they were being written on a blank slate. Water may be sold from federal projects at excessively low prices, stockmen may have indefensible advantages on federal grazing lands, inefficient timber management may be supporting marginal mills, but existing communities and industries have grown up around these practices.

Again, there is an irony here. Those who favor disposal tend to sympathize with the claims of property owners, and are wary of government-imposed disruption of their expectations. Should the expectations of those who have had long-standing benefits from traditional uses of the public domain be discounted simply because they are not—from a lawyer's perspective—formally cognizable property rights? There is a powerful tension here that suggests the imposition of prudential restraints on any program that threatens sharply destabilizing changes imposed over a very short period of time. It well may be that the result of the current system of political decision making, with all its inconsistencies so maddening to theorists, is more acutely sensitive to the real interests of the various constituencies that comprise the American public than appears to distant critics.

This leads to my final preliminary observation. Ownership, although it is the focus of the current debate over the future of the federal lands, is, in fact, a poor measure of the real relationship that exists between government control and private market decision making on the public lands. For example, nothing in the fact of government ownership itself prevents the government from managing its lands precisely as a private entrepreneur would do. Following the same model of behavior that the proprietor of an office building uses in leasing space in a skyscraper, the federal government could lease lands for mineral or timber production to maximize economic efficiency. By the same token, government could sell off all its lands and subsequently impose through regulation any of a variety of constraints that would impair the economically efficient use of the property. Both such models of behavior, and many in between, can be found somewhere in the vast recesses of the federal government. To those who yearn for economic efficiency as the governing principle, it need only be said that public ownership does not negate its possibility any more than private ownership in a regulatory milieu assures it.

It is often said that government bureaucrats are poor managers or that public ownership is excessively influenced by powerful interest groups, and there is evidence to support those observations. Selling the public lands will not cause those problems to disappear, for to the extent that the government as a proprietor is a source of mischief, there is little reason to believe that the government as regulator will be less mischievous.

In theory, there is very little constraint on private users of the public lands accomplished through public ownership that could not be accomplished by regulation of private owners. The scope of federal authority to regulate interstate commerce, for example, has been interpreted with great liberality by the courts, and there are very few regulatory schemes (other than management wholly for preservation in a natural state) the Congress might wish to enact that would not pass muster with the courts. Indeed, there is a great deal of privately owned land that is much more rigorously regulated at both the state and federal level than is most federally owned land. A great deal of public land has been made available commercially for timber harvesting, mining, hydrocarbon development, grazing, and entrepreneurial recreation (such as downhill skiing); and I think it fair to say that the private benefits from such uses have been considerable. This is perhaps just more reason explaining the absence of an intense groundswell of support for disposition, even on the part of private commercial and industrial users of the public domain. That lack of enthusiasm is especially true of small businessmen, like ranchers, in a time of high interest rates. Those who have been using the public lands (without having to bid for their use) can hardly be expected to welcome bidding against large corporations and foreign investors in order to continue their traditional uses.

My own speculation is that whatever popular support there has been recently for sale of the public domain was generated largely by fears engendered by the Carter administration's more aggressive public land-management policies. Commodity users saw the specter of increasing environmental regulation, greater restrictions on leasing, and increasing prices for such users as stockmen. To the extent that those fears were justified, sale to private owners might have been very advantageous. With the advent of the Reagan administration, many of those fears have been stilled. Insofar as the desire for privatization is a battle cry from a war that was won at the ballot box in 1980, it is hardly surprising to observe that it has lost much of its energy.

And what of those who oppose sale of the public domain? They would note that whatever the theoretical possibility of extensive regulation (for environmental protection, as an example) of private lands, it is much easier in practice to control lands in public ownership than it is to control those in private hands. It has been observed that many people who are enraged at the idea of federal zoning of private lands, or of extensive administrative regulation of those lands, are much less exercised when the same controls are imposed on publicly owned lands through contract or lease. Thus, sale opponents might say that ownership is important not in itself but simply because such concerns as ecosystem protection are in practice easier for government to implement as proprietor than as regulator. My own experience tends to confirm this observation, but it also suggests that the content of regulation depends a good deal more on the identity of the occupant of the White House than on the identity of the proprietor of the lands.

I conclude these preliminary observations with the thought that both proponents and opponents of sale overestimate the importance—*as such*—of ownership. I do not mean to suggest that ownership is never critical. Plainly certain kinds of uses, such as parks, weapons-testing sites, or wilderness, are extremely difficult to provide without proprietorship. But I do suggest that a profound controversy over the legitimacy and the importance of public values is often obscured by excessive focus on proprietorship.

The debate over ownership of the public lands is basically part of a much larger controversy over the legitimacy of collective versus individualistic values. In essence, the argument in principle for disposal of the public lands is this: Each person knows best what is best for him or her, and, therefore, the best system is one that permits the real preference of individuals to be revealed and implemented. With rare exceptions, the ideal mechanism for implementing these preferences is a private marketplace where each individual expresses his or her desires through bidding. Private ownership advances this goal, and public ownership impedes it. If, for example, I want to use land for hiking and you want to use it for timber harvesting, public ownership which sets it aside for hiking permits me to get what I want without paying its true cost, since the losses engendered by prohibiting timber harvesting will be spread throughout the tax-paying population and will not be borne by me. The outcome, unlike that which would be achieved by market bidding between the hikers and the loggers, will thus be imperfect.

It can be seen immediately that the problem thus suggested is not limited to ownership. It also is present in government regulation of privately owned land. If government regulates timber harvesting to advantage recreational users, the recreationists benefit without having to reveal their true preferences, that is, without being willing to pay for the losses generated by the regulation. It is not ownership alone that raises the problem of "distorted preferences," but control through government. If, it is sometimes asked, proponents of continued public ownership give as their reason the need to control externalities, why do they not urge the government to buy up all the McDonald's restaurants, for they are certainly generators of external harms with no less impact than many users of the public domain? The simple answer is that ownership and regulation are often used as alternative means of accomplishing the same end (one of which is to control external harm). Those who advocate the sale of public lands would not be any more content if the lands were sold into private ownership and immediately subjected to a wide range of regulatory controls, than they are with a continuation of public proprietorship.

The real issue that divides advocates of sale (or "marketeers") from those who seek retention (or "regulators") is found in the unstated assumption that underlies discussion of "preferences." The marketeers assume that the only real or legitimate preferences are those that are expressed by individuals behaving atomistically. Thus, if an outcome differs from that which would have occurred through the expression of individual preference (the sort that occurs in market transactions), *ipso facto* it must be wrong (except in those relatively rare instances where the market does not reveal true preferences). The regulators believe that individuals have more than one kind of preference, and that because individual behavior in the market reveals only one species of preference, it therefore is incomplete. There is, they say, a kind of preference that people hold solely in their capacity as members of collectivities, and for which only collectivities speak. One such collectivity is the political community, or the government. When the government regulates, or controls use as owner, it is expressing a collective preference.

Thus, for example, if the government restrains timber harvesting to the advantage of hikers, it is expressing a preference no less real or legitimate than the preference I express if I decide to keep my land as a rose garden rather than to sell it to a timber company at a given price. To be sure, that decision may differ from that which any individual or group, acting *as* individuals would have made, but that is simply because the political community as a collectivity has some interests which differ from those of any individual or group. The outcome is not "wrong" because it differs from that which individual action in the market would have produced. Nor is it wrong because somehow the collectivity does not "pay" for its preferences. If the government owns a tract of land and decides to forgo the potential benefits from timber harvesting in order to maintain recreational use, that decision is no different from the one I make in forgoing a large cash payment to maintain my rose garden. Of course, the way in which that willingness to pay is felt differs in a collectivity—it is more indirect, for example, and more diffuse—but that is the only way a collectivity can operate. To characterize its behavior as inappropriate, illegitimate, or inefficient is simply to deny the possibility of a collective value.

All that I shall say henceforth will expand upon the notion of collective preferences set out in the two preceding paragraphs. Perhaps a word about the terms we use is a useful way to open the discussion. I have thus far employed "preference" to refer to both individual and collective interests. One may object that it is misleading to so designate collective decisions by nonvoluntary entities such as the state. There may be a consensus expressed by such an entity, but it hardly can be called the preference of every member of the collectivity, including dissenters.

The word *is* an awkward one, and for more than one reason, but so also is the problem. If we start by saying that the implementation of preference is the only sure means to the appropriate maximization of both individual and social benefits, and we say also that preference only can describe individual, wholly voluntary choice, then we deny at the outset the possibility that collectively expressed choices can be legitimate, the very question I propose to discuss. Let us examine the following question, Why is it assumed that the only expressions worthy of consideration are those individually expressed by individuals acting atomistically, or by groups where the collective expression is simply the sum of voluntary individual desires?

Indeed, there is more to the use of preference than meets the eye. For one thing, the word usually implies indifference about the significance of the preference. We say, for example, you prefer blue shirts and I prefer white shirts. Preference has that connotation. Conversely, collectively expressed interests routinely imply values to which social significance is attached. To take but the simplest example, it would seem absurd to describe such collective statements as the Ten Commandments or the Bill of Rights as the Ten Preferences.

These semantic observations suggest that basically the preference/ individual/private market advocates are skeptical about the desirability and the legitimacy of most community values and of the imposed loyalty that they demand. To the extent that public landownership encourages the imposition of such values, such ownership simply becomes a particular example of a larger issue of social collectivism. Stated most starkly, the concern is that, as collective values increase, individual autonomy declines. And the corollary premise is that the maintenance of individual autonomy is the premier virtue, so important and so obvious that it barely needs to be defended. That premise often underlies the rhetorical question, What makes you think that the state or some bureaucrat knows better than I do what is best for me?

It is, however, far from clear that the maintenance of individual autonomy is the primary goal of most people most of the time. Indeed, there is a great deal of evidence to suggest that one of our strongest urges is to identify ourselves with a source of moral authority, and to subordinate our autonomy to it; that we draw strength from values external to our purely personal convictions; and that we draw values from collective solidarity. Psychological and anthropological studies confirm this observation,[2] and modern literature focuses on exactly this longing in its pathological extremes, suggesting just how powerful an element of personality it is. To illustrate, Thomas Mann's great short story "Mario and the Magician" contains the following passage: "The master had but to look at him, when this young man would fling himself back as though struck by lightning, place his hands rigidly at his sides, and fall into a state of military somnambulism, in which it was plain to any eye that he was open to the most absurd suggestion that might be made to him. He seemed quite content in his abject state, quite pleased to be relieved of the burden of voluntary choice."[3] Or consider this celebrated passage from Dostoevski's *The Brothers Karamazov*: "I tell thee that man is tormented by no greater anxiety than to find someone quickly to whom he can hand over that gift of freedom with which the ill-fated creature is born. But only one who can appease their conscience can take over their freedom."[4]

The individual who has yielded all autonomy is, as these passages reveal, only a slave. But they also suggest that the individual who is all autonomy is beyond the limits of human nature. Almost all of us, almost all the time, operate somewhere between these extremes, and it is within this intermediate position that such public controversies as landown-ership are likely to find their focus. For it would be difficult to locate a wholly autonomous individual anywhere in the normal range of be-havior we daily observe. Let me illustrate "normal" behavior with a commonplace example.

Imagine an average individual happily using a tract of land he owns as a rose garden. Another offers him $1,000 for the land, intending to use it to build a house. The owner makes a counteroffer to sell at $2,000, and settles with the builder for $1,500. This individual is operating in the realm of pure preference.

This average individual may well decide to give the tract to the church as an annual contribution—not an unprecedented thing to do. The church then refuses the highest offer made—$1,500—and determines to main-tain the rose garden as a senior citizen project. Assume, further, that the outcome differs from that which would have occurred in any indi-vidual transaction. The church, as an entity, has different "preferences" than the original owner, or any sum of church members, acting as in-dividuals, would have revealed. And assume finally that the original donor and his co-members of the church are perfectly happy with the result, though it diverges from what any of them would have done had they retained ownership of the property.

What has happened here? Why would someone give up autonomy over the land and permit it to be used this way? Why is it, in short, that people often choose not to express their interests individually, but through the medium of a collectivity?

One possibility is that some collectivities are administratively efficient units which, by summing the property of a number of owners, can more effectively achieve the result each individual wants than would have been possible were each participant to act individually.

Another possibility is that, since information is not costless, the col-lectivity can obtain information that the individual needs but cannot obtain. The collectivity, once informed, will make the decision each individual would have made if he or she had access to the information.

Still another possibility is that he or she wants the collectivity to help formulate his or her preferences. The individual might believe that he or she should do something about charitable contributions and, to this end, is willing to pay part of an advisor's salary in order to learn how much to contribute on an annual basis both for the present and in the future. A further possibility is that he or she endows some collectivities because of feelings of mutual obligation and duty and to that extent, wants the collectivity to go beyond the advisory role and to shape, intensify, or change preferences held previously. Certainly some collectivities, like churches, behave as if that were their mandate.

One might say that nothing very interesting is revealed by all this since such collectivities are voluntary, allowing any member to resign at any time that the organization does not fully reflect what is desired. Thus the collective decision, whatever it may be, is nothing more than the sum of the then preferences of its members; and the "different" decision of the collectivity is no different than the preferences an individual might express at two different points, depending on what information is available. Thus, the atomistic model still holds.

Yet—whatever the precise mechanism by which collectivities operate—the very decision to give resources to a specific collectivity has consequences. Even if the collectivity does nothing more than provide information that shapes the individual preference, it is not the same information that was available to the individual before he or she had made the endowment. The decision whether to resign subsequently is inevitably affected by the information, or advice, provided to the individual.

This is why action by "voluntary" collectivities cannot be dismissed as simply the implementation of the purely individual decision of each of the members, where the member's nonresignation is treated as a sort of subsequent ratification. The critical period is not the moment after the collective decision is made (at which time one decides whether to resign) but the period between entry into the organization and the moment the decision is made. That is, what happens up until the decision is made—which may be viewed as the information flow from the collectivity to the individual member—will itself change the person (by supplying new data) and will thus itself affect the decision whether to resign. At the very least, one who joins an organization yields autonomy in the sense that individual authority is transferred to another to obtain information that will inform subsequent decisions.

To give but one example of the shaping power of collective action, assume that a group of individuals get together voluntarily to establish an organization to meet their recreational needs, and that organization in turn establishes Central Park. After the park has existed for some time, each member then asks what he or she thinks about such parks. It seems inevitable that the individual's response will be affected by the mere fact that Central Park has existed up to the point that the evaluation takes place. It also will be affected by how much time passes before such an evaluation is made. Whatever the precise answer in a given case, the individual to some extent will have given up some autonomy to the collectivity, by letting the collectivity shape his or her knowledge and experience. And of course, letting *that* voluntary collectivity thus act on the individual is quite different from eschewing any such collective action, or letting some other collectivity (like the Catholic Church), which would behave quite differently, use some of one's resources, or preference power.

Of course, people yield to all kinds of collectivities, and the commitments they make vary enormously. Joining the Mormon Church is a much greater commitment (a greater yielding) than is subscribing to the *New York Times*, even though each group serves as a source of information and has a power to shape the individual "subscriber." The extent to which a collectivity departs from a model of purely autonomous behavior depends fundamentally on the nature of the organization and the kind of commitment that its adherents generally make.

What is striking is how willing so many people are to yield their autonomy in order to join and to remain in the collectivity. What this suggests is that maintenance of autonomy is not the only value, or the decisive value, in the everyday behavior of most people. However much people may value having their individual preferences implemented, they are also very interested in expressing themselves through collectivities, though doing so demands a price, and often a quite heavy one, in individualism.

Voluntarism clearly shows this. The fact remains that all voluntary organizations are not equally voluntary in the exit opportunities they offer, except at the most abstract theoretical level. Some collectivities so restrain the exit opportunities they offer that one is compelled to conclude that people would never join them if what they were seeking was the sort of free exit that is usually connoted by the term *voluntary*. People join them *despite* the fact that exit is very difficult, demonstrating that loss of autonomy (the essence of the atomistic model) is a price perceived as worth paying for the opportunity to participate in a collective identity as an important element of one's total identity.

If we wanted to ensure that use of our resources and our choices were always reflective of an individualized perspective, we would ensure the greatest ease of exit for all the collectivities we join. But that is plainly not the case. While some sorts of collectivities, such as business corporations, are obviously designed to assure the greatest ease of entry and exit to stockholders, others have precisely the opposite characteristics. How does one explain the commitment to lifetime loyalty and mutual obligation that so many churches have developed for their members in ceremonies that are administered to those born into the church at the time they attain the age of reason? Why, indeed, do ideas like patriotism and loyalty attain so much support in involuntary collectivities like the state, whereby we bind ourselves far more than we have to, to an involuntary collectivity, rather than simply affirming the minimal obligations and loss of autonomy essential to the maintenance of the nightwatchman state?

Consider, as a striking illustration, the way in which collectivities such as churches operate. Imagine an individual who is a member of the Catholic Church, and who makes contributions to it. If the church were functioning as a mere information-gathering, autonomous, and exit-maximizing entity, it would talk in the following terms: "We have studied the question of abortion and for the following reasons have come to the conclusion that abortion is immoral. We intend to use contributions to oppose abortion. Those who are in accord should stick with us; others should resign, or should remain in the church only for further information bulletins on moral issues that will be sent out from time to time."

In fact, the institution is structured to produce precisely the opposite reactions: To generate commitment and obligation, and to discourage resignation. Indeed, it is structured to encourage the assimilation of individually held views into a collective view in the name of what good Catholics believe. It encourages its members to have a position on abortion *as a Catholic*, rather than as an isolated individual whose source of information includes the Catholic Church.

Surely those people who associate themselves with such collectivities are aware that membership in these institutions is entangling, and that one does not cease being a member with as little psychic cost as one ceases to be an Exxon stockholder. Surely there is an awareness that with each dollar contributed, one contributes to the capacity of the institution to shape tastes and values in less than costlessly reversible ways. Does anyone who gives a child a religious education think that such a decision is fully described by saying, "Well, he can always resign." Or that the kind of society being created by people who behave like him is no different from one in which people maintain the maximum of personal autonomy to behave atomistically?

Yet, obviously such organizations as churches do not disappear, as would be expected if voluntarism, in the sense of ease of exit, were the critical element.

Let me make clear at this point that I suggest no absolute. I am far from suggesting that most people do not want any atomistic autonomy and want to submerge themselves totally in collectivities. Nor do I suggest that even the most tightly knit collectivities like churches are, or are desired to be, wholly involuntary, without possibility of exit, and without possibility of an interplay between the autonomous individual and collectivity. Indeed, I seek to make quite a different observation:

That there is a complex interplay, a mixture and a tension, between atomistic behavior and collective behavior, and that both kinds of behavior reflect normal elements of the human personality.

The import of these observations is that since people join collectivities and yield a good deal of autonomy to them, despite the psychic cost that resignation entails, they must join in part because they deem participation in collective expression important. If expression by the political collectivity—the state—is deemed important that tension is increased, since the state offers no right of resignation whatever. The political community is at the furthest extreme on the spectrum of voluntarism.

It is, of course, possible to believe that this difference must always be decisive; that the coercion the state imposes is always more harmful than the benefits that collective action by the state provides. But unless one believes this, then the question of collective action by the state is not one of legitimacy, but of prudence in balancing communal action against loss of autonomy. Is the game worth the candle?

Is one compelled to believe that such coercion is always to be avoided? I have already suggested that the observed experience with "not-so-voluntary" entities is some evidence against the proposition that retaining maximum autonomy is the premier virtue. But there is stronger evidence than that, for in organizing governments the public routinely (perhaps uniformly) yields to them more than the absolutely necessary minimum. The so-called night watchman, or minimal, state is in fact more a theoretical construct than a reality. Everywhere one looks, now and in the past, states take on functions such as the promotion of national solidarity by building monuments, supporting the arts, promoting historic consciousness, and advancing education. In short, communities build monuments to what they value, and history is replete with evidence that they often value commitments to solidarity more than mere single-minded devotion to individual autonomy. Perhaps this is only evidence of the sort that simply asserts the importance of something because people have found it important over long periods of time. But do those who think the maintenance of untrammeled autonomy is the central good to be achieved have any evidence *more* compelling to support their conclusion?

To assert the legitimacy, and even the desirability, of collective action by the state does not, of course, reject the importance of the individual; nor does it suggest that every collective value that can obtain a majority should, *ipso facto*, prevail over the claims of a dissenting individual. An intermediate position between the totally autonomous individual and the totally dominating state is what has most often commended itself to every society that takes account of the individual's importance. Rather than saying it is the duty of society to maximize personal autonomy to the greatest extent possible, they have said that *some* elements of personal autonomy are too important to be exposed to collective coercion. At least in the broad sense, those elements are eminently familiar: for example, freedom of speech, association, belief, and religion.

The distinction thus made between carving out a range of personal rights immune from collective coercion and prohibiting coercion everywhere except in a very few essential areas (such as taxing for a police force) serves as a recognition that (1) maximization of autonomy of every kind is not the sole or always dominant social value; (2) that expression of collective values is very important, even though some autonomy is thereby sacrificed; and (3) that the collective values of the political community can be deemed essential, even though the community is wholly involuntary.

This conclusion—which roughly describes the standard American position—as well as that of almost all political communities that share essentially in the same philosophical tradition, is significant in recognizing more than one kind of interest as both legitimate and important, despite the tensions that exist among them. It thus adopts a posture that does not embrace either value (collective expression or maximum autonomy) as of dominating and exclusive importance.

If the intermediate solution I have just described makes any sense, it must rest upon a conviction that there *are* collective values both important and distinctive to the political community. For, if there are not, there is no justification for requiring any citizen to yield any autonomy to the state.

I suggest that there plainly are such values, which I shall try to illustrate with the following example. Imagine a country resembling France having a strong sense of nationhood and a strong interest in what the people call their patrimony. Suppose the country has a large number of historic châteaus—oozing with historical importance and greatly in need of restoration—situated at places that make them prime targets for commercial development. There are a considerable number of people who are interested in both visiting and preserving the châteaus, and though they are willing voluntarily to pay a good deal to purchase them, in general they cannot outbid commercial interests.

Deeply discouraged, they are ready to abandon the effort, believing that their bids represent the true value of preservation, which is, lamentably, less than the value of the châteaus for commercialization, when one of their number makes the following observation.

The interest in preserving these châteaus has been miscalculated up to this point. The preservation value consists not only in the opportunity afforded to us (and others like us) who will visit the places, and who obtain gratification from such preservation; but the entire nation will benefit if patriotism is affirmed and a sense of nationhood engendered and sustained by the maintenance of these buildings. There is no reason why we should pay, or be expected to pay, for values beyond those which flow to us directly because of our special interests. Nor should we pay for those which flow to us as part (but not the whole) of the nation that will benefit by increased national solidarity. We have lost out in the bidding because the true value of preservation is greater than what we bid, or what (acting as individuals) we should bid. It is perfectly possible that if the nation itself bid for the properties, it would outbid the commercial developers.

Still, one may ask, are we not able to generate voluntary contributions that equal the totality of these benefits? The answer is clearly no, for some well-recognized reasons. If one of the benefits of preservation is the strengthening of national solidarity, then preservation will automatically benefit every citizen. Even those who do not contribute voluntarily will nonetheless reap the benefits; this is what is familiarly called the "free-rider" problem. Moreover, *this* benefit (as contrasted to the benefits we reap as individual visitors, for example) will not exist unless and until the nation acting *as* a nation recognizes it. That is, there are some values that become important to me, as a citizen or as a communicant of a church, only on the condition that all my fellow members join me. I may, for example, be more willing to contribute to charity to sustain the poor if everyone else joins me in that goal, because it is not only the distribution of benefits that are important, but also the perception of the poor that the community cares about them. The reason for this may stem from a belief that if the poor are taken care of they will constitute a less-disaffected and revolution-prone class. I can only achieve that goal (like the goal of national solidarity) if there is a community-wide commitment to it, denoted by an "official" act of the whole community. My interest as a member of the community may be perfectly selfish, but it nonetheless is distinct from my interest as an individual.

The situation just described is, I admit, a bit complicated. Something becomes a value to everyone in the community only if the community, acting as such, affirms it as such. It is not less a community act even though there are dissenters; and it would not be a community act if the same number of dissenters absented themselves from a voluntary decision about the same subject. It is rather like a situation where the Pope is recognized as Pope by an act of the Church, even though it is not unanimous. Whereas the notion of a Pope recognized only by those Catholics who wish to recognize him explodes the very idea of a papacy.

The mere fact that the political community adopts a value does not make it a desirable value. Nor does the fact that a community-adopted value refrains from intruding on fundamental individual rights make it desirable. That leaves the question, should anything that can garner a 51 percent vote in the legislature be blandly accepted as a statement of the collective values of the political community? My example of the châteaus was designed to show that there are some collective values (such as preservation of the nation's history) that are widely accepted as legitimate and desirable even though they go far beyond the goals of the nightwatchman state, values that do not intrude on what most people consider fundamental and ones that differ from those that any group of individuals would have, acting as individuals. That is, they are "existence" values of a sort that promote national feelings of solidarity, over and above whatever "consumptive" values they have to users, or "existence" values they have to history buffs.

The question remains whether all majoritarian acts of the state are self-validating as desirable simply because they *are* the majoritarian acts of the moment? Surely not. Only a fool would suggest that everything the Congress does is sensible any more than it could be suggested that everything Joe Doakes or some business does (like the manufacturer of Edsels or thalidomide) is sensible.

One might believe, as some do, that, in general, government does more mischief than good and that, in general, private interests do the opposite. At one level, there is no arguing with one's perception of the reality perceived by an individual, but it may be useful to disaggregate these perceptions as much as possible. At one level, the argument is that individuals and businesses are more disciplined because they pay for their mistakes. By and large, they are in business to make money and, if they fail, they will be thrown out.

This observation calls for several responses. One is that the goal of the individual or entrepreneur in the market is, generally, more straightforward than the goals of the state, or of most collectivities. It is easy to test whether one who is trying to make a profit is succeeding. It is a great deal more difficult to test whether one who is trying to advance charitable goals or encourage morality (like a church), or promote education (like a school), or strengthen national solidarity (like the hypothetical state described above) is succeeding. Indeed, one might point out the one governmental role that almost everyone accepts—promoting national defense—and note how much controversy that activity generates. Are we succeeding in making the country more secure by throwing more money at the problem? Are we failing because we try to prop up weak dictatorships? Would a worldwide income redistribution do more to make us secure than continuing the arms race? There are no easy answers and no easy tests of success.

Similarly, there is thus no objective test as to whether we are sacrificing too much to save the grizzly bear, any more than there is a clear answer (either for an individual or for a public entity) as to whether it is spending too much for education or for the support of art museums. And there is no obvious way to know whether either grizzly bears or art are worth saving. It should, however, be noted that even governments are not willing to spend endlessly (or endlessly to forgo other benefits). Its values may be different, and it may be more forebearing than individuals because its losses are less immediately evident (as critics never tire of pointing out).

But it is important to observe that not even government is immune to the claims of competing values, including those of economic efficiency and material well-being. Whether or not one thinks the government has designated too much wilderness, it can hardly be denied that those who would like to use that land for other purposes, such as hydrocarbon development or mineral mining, have ample opportunity to make their case in the Congress. And it could hardly be said that Congress is deaf to commercial and industrial interest groups or that it is insensitive either to the economic benefits it could reap from commercial activity or to the implications of such choices on questions like tax rates.

Another complaint frequently made by "marketeers" is what might broadly be called the "corruption" of the governmental process. It is suggested, for example, that bureaucrats have dominant control over public programs and that they act selfishly to promote their own well-being rather than the public interest. I do not see how anyone could deny the existence of such a problem, but it must be recognized that this problem exists with any organization employing intermediaries. The tension between corporate managers (who have interests of their own) and stockholders is a staple of the business literature. Every private, collective entity—the churches, the universities, labor unions—are prey to this problem.

It also has been emphasized that the state may be, and often is, at the mercy of powerful interest groups who turn government to their own limited interests. Again, there is no doubt that this is a problem, and a serious one. It is likewise true that elected officials are far from perfectly accountable to their constituents. Some are fond of noting how few Americans even know the name of their representative in Congress, and even if they do, how difficult it is to make a periodic vote reflect any constituent's view of the legislator's vote on the myriad of issues on which he will express himself in the interim between elections. Of course, all these are common, and familiar, failures of politics. No sensible person denies them; no knowledgeable person can dispute them. It is not the existence of the problem that is in controversy, but the solution suggested.

The answer in effect given by the marketeers is to give up on politics, to view it as irredeemable. The suggested response is abolition, rather than reform; or, to put it rather more modestly, to remove from government as much power as possible to do mischief, by limiting its functions to the minimum possible level. Again, I concede that this is a legitimate philosophical view, but I must emphasize that it incorporates a very controversial assumption: That most of the things government seeks to do—the sort of things that are distinctively the product of collective values held by the political community—are relatively unimportant. Thus, in the view of these critics, to give up most collective values is to lose little or nothing. The critique reduces to the proposition that only those values which we hold as individuals, acting atomistically, are worthy values. There simply *is* nothing beyond personal preference. If one is persuaded by such a view, of course there is no point in working at reform of public decision making. My goal here has been to suggest that the assumptions underlying such a position are far from obvious.

Having said this, let me return to the question of whether there is any standard by which the propriety of public collective values, or preferences, can be judged. The answer is both yes and no. Yes, of course, at a moral and ethical level, collective values can be judged just as personal preferences can be. At a less profound level, however, I hope I have shown that they cannot be judged simply by measuring them against the preferences that individuals, acting atomistically, would have revealed.

What the content of such collective values or preferences should be, however, outside of the profoundest moral judgments, cannot be identified any more than the values of an individual. A community (or an individual) may invest with a high value the efficient production of commodities, and may value very little the preservation of a grizzly bear or the provision of public recreation. Or it may adhere to precisely the opposite priorities. At various times, people have held widely divergent views. At one time there was thought to be no value in pornography; today it is seen as in deep tension with the values of the First Amendment. Abortion, once the subject of almost universal consensus, has become a deeply divisive issue. Free public education was not always thought fundamentally important; public parks are a rather recent invention, though today they are widely considered indispensible. One could extend the list of changing public values almost endlessly.

If there is to be a place for the expression of civic values, the only means available (unhappily, perhaps) is through the decision-making apparatus of the state. That the state is imperfect and corrupted in many respects (as are churches and labor unions) is an inevitable price of representing the interests of the political community. That the expression of political community values involves even more coercion than do collectivities from which resignation is possible makes the price even higher, though not necessarily intolerably high. Whether the interests are important enough to bear the price of corruption, and whether efforts at reform are worthwhile, is a question on which no indisputable statements can be made. I respect the judgments of skeptics on this question, although I do not share them. I do insist, however, that they *are* judgments only and nothing more.

All this brings me finally to the specific question which is the subject of this conference. What possible distinctive interest could the national political community have in the hundreds of million acres of publicly owned land, mostly in the western United States, and largely valuable primarily for commodity production, that would induce it to maintain ownership? I must begin by repeating what I said at the beginning of my paper—that ownership *per se* is not the question, although it may have great operative significance in certain settings. The real question is whether there is an interest in *control*, in imposing some constraints that purely private decision making would not engender. If there is, then it may well be as a practical matter that ownership (rather than sale, and subsequent regulation) is the most likely means to effectuate those controls. I might note also that retaining ownership does not necessarily mean retaining total control. The government might permit private uses, and the goals of the private market, to dominate the bulk of use of the land, retaining just a modicum of control in the form of public values (free public access for recreation, or constraints on mining in certain sensitive areas).

I think the answer to the question of the political community's possible interest in the public domain lies in a much more general relationship between public values and the use of land. Without hoping to be exhaustive, let me sketch briefly the transition I see taking place in that relationship, of which the debate over the public domain is only one modest element.

I must begin with the trite observation that there is always some link between the rights that individuals are permitted to obtain in property and some public notion of the public interest. Perhaps the point is most easily illustrated by a very old issue, the right of inheritance. At the outset it is always the political community that decides what interests people may, and may not, obtain in property. To permit individuals to acquire the right to transmit their property to their heirs through inheritance demonstrates a social value. Laws that prohibit or impair the transmission of wealth through inheritance embody quite different collective values. Every rule of property, including rules about what rights of use, or sale, or inheritance, are adopted, begins with some social value.

Of course the conventional rules about property in America are socially based too, though these original social values are now so deeply embedded in traditional thinking that we tend to accept them as inevitable. Our tradition, made explicit by writers like Blackstone, views it as desirable that private rights be given out very extensively, and that private owners should be permitted to do as they wish with their property, conditioned only by the constraint that they should not cause affirmative harm to others (such as creating a nuisance).

That is a view of private ownership perfectly consistent with a production-oriented, developing, and industrializing world. For the kind of uses owners would generally make of their property when left largely free were the sort of productive, developmental uses that the society viewed as progress. So that, generally, what was good for the owner in producing profit and personal benefit was also perceived as being good for the society in producing economic growth. One might make the same point about property and the "cowboy" society of the American West in earlier years. Both the settlement and taming of the West, and the development of its mineral and other natural resources, were in harmony with American political policy for the West.

Thus, traditional views of the rights that property owners had, and should have, were not "made in heaven," but were compatible with, and grew out of, the dominant collective values of the time.

Today we are in a state of transition away from some of those values. I do not wish to be understood as overstating the point. I do not assert that conventional ideas of progress and economic growth are dead; far from it, I assert only that those ideas are less dominant and unanimous than they once were and that there is much more controversy about those traditional values than there once was. The so-called environmental movement, to take but a single example, is one instance of a growing (though perhaps still a minority) view that growth is not so good, or at least not unquestioningly good, or unlimitedly good.

As traditional values feel the stress of change, the institutions that grew out of those values will be reexamined, as, indeed, they have been.

One might also expect significant changes regarding a shifting relationship between the rights of owners and the scope of public regulation. And we have seen—particularly in recent decades—some very dramatic changes.

To put the matter simply, we have seen increasing regulation constraining traditional ownership—regulation that implements skepticism about the desirability of largely untrammeled economic growth. The most obvious examples, of course, are conventional air and water pollution laws, constraints on hazardous substances like pesticides, and controls over the management and disposition of wastes. All these laws are, as industrialists are fond of pointing out, limitations on economic development. So they are. It might be said that these laws are simply modern examples of the sort of control over harmful externalities that can be traced back centuries to the law of nuisance. So they can, though the balance of interests has certainly shifted away from the traditional encouragement of industrialization that made nuisance a quite limited legal remedy.

But there is also a wide range of controls on property that have no obvious antecedents and which underline sharply the growing idea that nonuse, or preservation, rather than development and exploitation, may well be the highest and best use of property. One example is the recent growth of historic preservation laws, which are remarkable when considered from a traditional perspective. Owners who have done nothing "wrong," who cannot in any ordinary way be said to be imposing harm on others, are nonetheless often required to leave their property as it is, because it is believed that retention of historical structures is a more valuable use of the property than any developmental changes would be. Suburban growth-control ordinances are another example of the same phenomenon; however controversial such laws may be, they are revealing of a sharply changed notion. Communities that traditionally encouraged development, and measured their success by how rapidly they were growing, now often seek to slow or even end growth, because they view maintenance of their rural character, or their quietude, as of the highest importance. Open-space ordinances in towns, and wilderness designation in the country, illustrate a similar principle: Doing nothing is viewed as the highest and best use of the land.

These are only a few of the best-known examples of the kind of change that is occurring to some degree everywhere in America. The change reveals itself in increased regulation because the "new value" is not one that is likely to be implemented by private owners pursuing their own interests. This is simply one more way of saying that traditional private property rights (which will lead to development and use) are becoming increasingly divergent from public values about property. Development and use, to the extent that it fails to promote what is viewed as "progress," sets the public and the private owner on different courses. The likely result is a change in the definition and content of property. And that is what the above examples all illustrate.

From the perspective I have just identified, the controversy over the public lands begins to come into focus. There is resistance to sale, because public ownership is seen as a means (and perhaps the most effective means) of control. Control is seen as necessary—more necessary than ever—because public values are more than ever divergent from the interests of private owners. Private uses are still thought appropriate, of course, and they may and often will still dominate the public lands. But those uses are increasingly constrained, just as urban uses are constrained, by zoning, growth control, environmental legislation, historic preservation, architectural controls, open-space regulation, and a host of other elements of the "new idea of progress." Retention of the public lands is really just another version of what occurs by regulation elsewhere.

To retain public lands only where they now exist—principally in the West—and to leave the rest of the country in private ownership is not as "inconsistent" as it may seem at first glance. It is control in favor of civic values—rather than proprietorship—that counts; and that control is moving toward a national equilibrium, as I have just noted above. In practice, there are doubtless some sensible reasons to treat the public land areas differently. They are, by and large, less developed already and are thus more likely to be affected by greater regulation than is the urbanized East; people in the rural West are less accepting of regulation by zoning, and thus—as I noted earlier—regulation by ownership may be a more effective way of achieving the same goals; some public values—such as wilderness preservation, wildlife habitat, and "open" public recreation—can only be pursued effectively in the West, and such goals are most likely to conflict with conventional private ownership.

The important point, however, is not proprietorship as such, whether in the West or the East, but the conflict between collective and private values, for the movement to sell the public domain lands is ultimately a symbolic one. From that perspective, sale of the lands (assuming it would lead to less regulation than now exists, a disputable assumption but not an obviously erroneous one, especially with the tone that has been set by the current administration in Washington) would in fact be a step backwards from the new balance that is everywhere being struck between the claims of privatism and of public values.

Notes

1. *United States* v. *Gettysburg Elec. Ry. Co.*, 160 U.S. 668 (1896).

2. For example, Sigmund Freud, in James Strachery, ed., *Group Psychology and the Analysis of the Ego* (New York, W. W. Norton, 1959) p. 24.

3. Thomas Mann, *Marco and the Magician, Stories of Three Decades* (New York, Alfred A. Knopf, 1936) p. 560.

4. F. M. Dostoevski, *The Brothers Karamazov* (New York, Modern Library, 1950) p. 302.

7
Weaknesses in the Case for Retention

RICHARD L. STROUP

Professor Sax's paper is a valuable exposition of several important public issues regarding the sale of federal lands into the private sector. It is important to note, as Professor Sax did, that the issues he discusses have little bearing on the current Asset Management Program of the federal government. Instead, he is talking about what is commonly known as "privatization," a large-scale divestiture program proposed by academics. By contrast, the Asset Management Program is a small, careful program involving land sales following extensive planning and public involvement. Our discussion relates to the academic proposal, and has only a limited relevance to current policy.

Points of Qualified Agreement

Sax makes the point that the appraisal of land is an inexact art. Indeed it is. No one person can be expected to know which use or which user can, in fact, make the best use of a piece of land in the future. Yet future use is what determines the price of an asset such as land. That is exactly why auctions often are held, and why the first offer is seldom accepted when a piece of property is up for sale. What sellers normally do is let all potential buyers have a chance to come forward and express their views, in the most sincere and concrete form: an offer to buy backed up by a willingness to pay. It is precisely the lack of centralized knowledge and, indeed, its infeasibility that makes efficient central planning virtually impossible. On the other hand, to capture fair market value, one only has to conduct a competent auction. To sell at such an auction is to capture not the average evaluation, but the high bidder's best offer for what is sold.

Sax is certainly correct in stating that a person's experience and association with others can change that person. Only with private, transferable ownership, however, are people free to associate with other potential owners and bid upon whatever strikes their fancy. All partnerships, the Audubon Society, religious groups, and a great many communes nationwide do exactly that. They need not seek the approval of, or action from, a government controlled by a majority of the population, who may be steadfastly opposed to the ideas of a smaller group. Indeed, as an association itself changes, it can sell or exchange its private holdings accordingly at any time it wishes. Such flexibility is difficult within a government setting.

Correct also is Sax's assertion that government is often at the mercy of powerful interest groups, and that it is corrupt inevitably, "as are churches and labor unions". The big difference however, is that one can leave a corrupt union or a corrupt church. It is far more difficult to leave a corrupt federal government.

Finally, I must agree with Sax that there has been a transition of values in our society toward more preservation of natural values. Both government and the private sector have recognized that trend and have acted accordingly. For example, in our Department of the Interior, the BLM—once known jocularly as "Bureau of Livestock and Mining"—has taken on many conservationist goals and preservationist personnel. Likewise in the private sector, recreation developments which previously have been aimed at providing easy access and many of the comforts of home are now being developed with far more emphasis on preservation of the natural features and differentiation from civilization, features which drew people to the developments in the first place. For example, in my home state of Montana, one can contrast the older development of West Yellowstone (which to this day calls itself "Snowmobile Capital of the World") with the newer nearby development of Big Sky. Developers of the latter first bought and traded to obtain virtually all of the land in a large mountain valley, in order that its neighbors could not put cheap and garish commercial installations near the carefully preserved natural beauty of the Big Sky development itself. After planning was accomplished, Big Sky sold off many small parts of the development, but under strict legal covenants. With virtually no government interference, neon signs were prohibited, architectural standards were set; each piece of land sold for correspondingly more money. The new conservation ethic and esthetic required a new way of doing business; profit-seeking entrepreneurs quickly responded. Posh accommodations, budget-level facilities, and spartan hostels were all provided in a carefully preserved setting. The smart money catered to the changing tastes.

Points of Disagreement

Although Professor Sax makes several good points in his paper, there are others with which I thoroughly disagree. For example, he claims that a collectivity will be better informed than an individual. Both logic and the facts disprove this statement when it comes to specific cases or projects. Mountains of statistics do not substitute well for the key facts and intuitions usually developed by owners whose personal wealth is at stake.[1] Since an individual is seldom decisive in a collective setting, he has little incentive to be well informed. Even if more information would change his vote, his vote would almost never influence the outcome. As a result, voters are seldom well informed. For example, only 46 percent of all Americans who meet the qualifications to be a voter can even name their congressional representative.[2] The electorate, while quite intelligent, is simply uninformed. The average voter is ignorant in a very intelligent way: time and information-gathering energies are allocated where they will make a difference *to that voter*. Let us consider an analogous case.

The average farmer who is trying to maximize his income will not pay much attention to studying meteorology, even though in any given year his income will be determined in large part by the weather. He remains rationally ignorant about meteorology simply because extra knowledge on his part will not influence the outcome. All his studying would not help him change the weather. Similarly, the average voter realizes that added time and effort spent at studying political issues can seldom lead to personal gain for that voter. The unfortunate result is that collective choices made in a representative democracy (the best device I can imagine for collective choices) are all too frequently made more on the basis of special interest pressure than on the basis of "the public interest" and "scientific management."

"The government" does not decide anything, despite what Sax and others have indicated. There often is little relationship between decisions made by individuals in the political sector, and how people would vote if there were a referendum. An analysis of legislative and referendum outcomes by Gwartney and Silberman indicates a specific pattern: concentrated interests win in state legislatures regardless of referendum outcomes on the same issues in the same states.[3] We simply cannot expect elected officials to seek or to receive, much less to carry out, the carefully considered wishes of the general public in managing the public lands. Instead, organized special interests will dominate.

Contrary to what Sax claims, I will argue that the public should not generally speculate on rising resource values. Why? How often, on average, will today's most optimistic market bid for a resource turn out to be too low? Work by Barnett and Morse, and others at RFF suggests that over the decades, prices of raw resources have generally fallen in real terms.[4] Extraction and recycling technologies, and added product substitutability have combined to overpower what some might believe to be the inexorable rise of scarcity. The fact is that resource prices, in general, like stock prices, generally take a random walk. Unforseen technological advances periodically reduce the demand, or increase the effective supply, of resources.

Another related point has been missed entirely by Sax: private speculators provide the only link that future citizens have with resource decisions in the present. Presently, as voters, future citizens have no clout at all. Politicians who hurt today's voters to help those of tomorrow harm their election-day prospects. That simply is not a survival trait for politicians. But speculators who correctly forecast price trends can buy today, taking off the current market a resource (development land, for example), and preserving it for the future. Eventually they will gain by selling it when others have begun to recognize the increased scarcity, and the price increases. The speculator who is even one or two years ahead of the market in his knowledge or expectations can cash in as soon as others jump in, by buying claims to that asset. Note also that a similarly farsighted BLM director or secretary of the interior cannot gain personally by guessing right and "swimming against the tide," sacrificing current benefits to preserve when most people do not see the reason for it. Further, only the generosity of voters permits preservation, even when the majority knows it is good for the future. With private property, charitable instincts are reinforced by future gains in property values, as both speculators and conservation organizations have realized.[5]

Sax indicates that to sell off land is to "shirk environmental responsibilities." Precisely the opposite is usually the case. A spectacular recent example is that of the Love Canal. Most people know that the Love Canal was used by Hooker Chemical Company as a chemical waste dump. Some even realize that Hooker was very careful to line the canal in a sophisticated fashion before it put the chemicals in. It did so in part to avoid legal liability from leakage. However those carefully constructed canal walls later were caused to leak. What few people realize is that Hooker did not willingly sell off the covered-over canal, and did not sell to developers. Instead, they were forced to sell the land to local government, under threat of condemnation. They did so at a price of $1.00, publicly expressing much concern over its later use. It was that local government which later built a school and sold to developers the remainder, and which approved the construction of sewers, the excavation for which pierced the canal walls. That construction, directly violating the earlier pleas of Hooker, was what allowed the chemical wastes to escape. Hooker Chemical had been horrified by the potential problems and its own financial liability, should just such an eventuality occur. Hooker wanted no development on the site. Government decision makers had no such liability, however, and the problem was quickly forgotten once the property fell into government hands.

It is important to recognize the twin forces which are present in private property ownership, both pushing strongly for responsible environmental behavior. First, Are the greater land values (and thus private wealth) associated with esthetic beauty, productivity, and a lack of environmental danger? Second, Is the ever-present threat to a private owner of legal liability, should environmental responsibilities be ignored? These forces are not present for the public decision maker. Only a highly ignorant electorate and the public decision maker's own conscience stand between us and serious environmental degradation when the land is owned publicly. Many have documented the environmental and fiscal atrocities committed by well-intended, but misguided, public servants unconstrained by market forces and legal liabilities.[6]

Professor Sax seems to believe that the rich benefit less from public ownership, and the poor benefit more. However, if we look at those who benefit from low-cost government grazing, the environmentally destructive practice of chaining, Rocky Mountain logging, expanded national parks, and the preservation of amenities appreciated most by those in upper-income brackets, one realizes that the poor are woefully underrepresented among the beneficiaries of such programs. The poor are seldom in a position to take advantage of underpriced grazing. What evidence we have suggests that park users are well above average in income, even though they pay on average about a nickel per visit to the parks. It is no accident that *Audubon* magazine advertises Rolex watches and Glenlivet scotch, rather than Timex and Budweiser. The advertisers know the audience for the magazines published by organized environmentalists.[7]

Contrary to Sax, it is true that government ownership does tend systematically to negate economic efficiency. This is not because government decision makers are less able or less well intentioned than any of their individual private sector counterparts. Instead, it follows from three major factors: first, the critical information cannot be collected centrally, since it is impossible for those who would transmit such knowledge even to know what is the important data to be transmitted (one might call this the "Hayek Problem"). Second, there is the problem of dealing with special interests: namely, the political system needs their support but cannot afford simple and obvious subsidies to them, but rather must hide their subsidies in circuitous and inefficient transfer mechanisms. Third, there is what Ann Kreuger has called "rent seeking," as epitomized by lobbying: when government favors are available, there will be destructive (expensive) competition among those seeking such favors.

Sax also brings in two perennial bogey men—"large corporations and foreign investors." It is easy to suggest that the sale of public lands will lead to ownership by these two "villians." Of course when one looks at the state of Iowa, and the state of Maine, both of which are almost entirely private, one does not note domination of land ownership by either. Then, too, when large corporations do have large landholdings, as does happen in Maine, their lands typically are easily accessible to the rest of us commoners. In any case, although foreign investors or large corporations could certainly find a way to buy virtually any or all of either state, such an action has not taken place.

Sax appears confused at some points. For example, he states, "When the government regulates, or controls use as an owner, it is expressing a collective preference." He seems to approve. However, later he admits that "the mere fact that the political community adopts a value does not make it a desirable value." Since he does admit that the state is particularly at the mercy of special interests, one wonders about the legitimacy or desirability of these collective decisions. Tobacco subsidies, the Garrison Diversion project in North Dakota, and the federal program of chaining millions of acres in the Piñon–Juniper Forests, mentioned above, are easy examples.[8]

Although Sax recognizes correctly that there are external effects in the private market, so that parties who are neither buyers nor sellers may be helped or hurt by economic activity, and that these effects are not fully taken into account in market transactions, he fails totally to recognize that external effects are virtually pervasive in government. It is almost never true that a government decision maker is held fully accountable for his decision in the sense that his personal gain or loss is proportional to the social gain or loss. In the real world of ignorant voters, a more imperfect connection could hardly be imagined. The Love Canal incident outlined above is but one graphic example.

Conclusion

The "collective interest" of Sax is taken by him to be the environmental interest, which he suggests should always outweigh development interest. This is just as unwise as what I perceive to be the old (circa 1955) Chamber of Commerce "boomer" approach—more commerce, more money flow, is better. Neither of these is very attractive. By making some institutional adjustments, including better control of externalities, and with the cautious sale of some federal lands of the sort the administration has begun, we will be able to reduce other environmental problems and simultaneously increase economic benefits from public lands. This we should all prefer.

Notes

1. See Thomas Sowell, *Knowledge and Decisions* (New York, Basic Books, 1980) p. 13.

2. Louis Harris poll, conducted for the U.S. Senate, Committee on Governmental Operations; published as *Confidence and Concern: Citizens View American Government* (Washington, D.C., Government Printing Office, 1973) pp. 215 and 216.

3. See J. Gwartney and J. Silberman, "Distribution of Costs and Benefits and the Significance of Collective Decision Rules," *Social Science Quarterly* (December 1973) pp. 568–578.

4. See V. Kerry Smith, ed., *Scarcity and Growth Reconsidered* (Baltimore, Md., Johns Hopkins University Press for Resources for the Future, 1979); and Harold J. Barnett and Chandler Morse, *Scarcity and Growth: The Economics of Natural Resource Availability* (Baltimore, Md., Johns Hopkins University Press for Resources for the Future, 1963).

5. More information on this point can be found in Richard L. Stroup and John Baden, *National Resources: Bureaucratic Myths and Environmental Management* (San Francisco, Calif., Pacific Institute, 1983) pp. 24 and 25.

6. See, for example, John Baden and Richard L. Stroup, eds., *Bureaucracy vs. the Environment* (Ann Arbor, University of Michigan Press, 1979).

7. On this point, see William Tucker, "Environmentalism and the Leisure Class," *Harpers* (December 1977) pp. 49–80; H. J. Vaux, Jr., "The Distribution of Income Among Wilderness Users," *Journal of Leisure Research* vol. 7, no. 1 (1975) pp. 29–37; and Joseph Harry, Richard Gale, and John Hendee, "Conservation: An Upper-Middle Class Social Movement," *Journal of Leisure Research* vol. 1, no. 4 (1969) pp. 247–254.

8. See Baden and Stroup, *Bureaucracy*.

8

The Case for Divestiture

B. DELWORTH GARDNER

Not for a half-century at least has there been so much discussion of the question of ownership of the public lands. Five years ago, many western states actively promoted a state takeover of the federal lands for a host of reasons, the most important being (1) a frustration with federal policies that more and more removed decisions on land use from influence by regional commodity users, such as ranchers, miners, loggers, and irrigators; (2) a paralysis of policy, where even rather trivial resource allocation decisions, to say nothing of important ones, were often delayed for years in court actions and bureaucratic red tape; and (3) a power struggle over who would capture the large economic rents that would be available from resource use, especially from the development of subsurface wealth in hardrock minerals and energy fuels.

Much of this initiative for state ownership dissipated when a conservative westerner was elected president on a strong states' rights, New Federalism, pro-development platform. It was widely perceived in the West that the pro-environment, antidevelopment policies of the Carter years would be abruptly reversed by President Reagan and then Secretary of the Interior Watt. Perhaps the water, grazing, and recreational

Author's note: Some of the material in this paper was adapted from the author's paper, "An Economist's View of the Production and Allocation of Products from the Public Lands," presented at a forum on "Natural Resource Economics: New Approaches for the 80's," which was sponsored by Montana State University and the Center for Political Economy and Natural Resources, and held in Bozeman on February 15, 1983.

156

subsidies and the sharing of public land revenues with the states connected with federal ownership and management were not so bad after
all. If the Sagebrush Rebellion is not dead, it is certainly fast asleep—
perhaps never to be reawakened.

Instead, the focus seems to have shifted to the mismanagement of the
public lands, and as to whether it is serious. If it is, what is to be done
about it? Some have argued that management reform directed toward
greater efficiency is what is required. The most compelling challenges
to inefficient federal management have come from libertarian and public
choice economists who have argued that mismanagement can only be
eliminated if decisions are placed in the private hands of efficient utility-
and profit-maximizers. While this might be partially accomplished by
long-term leasing and the insertion of various market processes in re-
source-use decisions, the simplest and most effective cure for current
mismanagement would be federal divestiture of the public lands into
private ownership.

The president himself gave that cause some impetus when he endorsed
the privatization concept in his fiscal 1983 budget message.[1] He got
serious about it in Executive Order 12348, issued on February 25, 1982,
wherein a Property Review Board was established in the Executive
Office of the President. It is clear, however, that he was thinking of a
very limited program of divestiture, since he said in his budget message,
"We will move systematically to reduce the cost of holding surplus land
and real property, [since] some of this property is not in use and would
be of greater value to society if transferred to the private sector. In the
next three years we would save $9 billion by shedding these unnecessary
properties while fully protecting and preserving our national parks, forests, wilderness and scenic areas."[2]

No doubt this action by the president encouraged those who had been
advocating disposition of *all* the federal lands, including the forest, parks,
and wilderness lands.[3] At the same time, there was a huge public outcry
by environmentalists and liberal columnists decrying any sales of public
land at all as they reminded the public of Teapot Dome and other
giveaway scandals of another generation. The public furor caused the
Department of the Interior to back off, "privatization" was discarded
as a term, and the administration's program of limited land sales became
known by the innocuous appelation, asset management program (see
chapter 7).

It is too bad in a way. In my view a respectable intellectual case can
be made for privatization of the public lands just as a case can be made
for their retention. It is a pity that the climate apparently is not conducive
to a rational public debate of the issue without the invective that emerges

whenever the subject is raised. However, we shall bear that criticism and press on with a public discussion of the issues in the hope that a rational public policy can emerge.

Perhaps the place to begin the case for divestiture is to state quite explicitly that in this chapter ideology will be avoided as much as possible, although "ideological" freedom of action is not a social value to be dismissed lightly. Rather, the alternatives of retention and divestiture will be primarily evaluated in terms of their economic efficiency, although distributional fairness will also enter the discussion occasionally. Some think even "economic efficiency" is ideology, but I prefer to think of it as a scientific paradigm. I take this ground because of an ethos in contemporary America that efficiency and fairness are significant, if not predominant, social goals and will serve well as criteria by which alternative policies can be evaluated.

Economic efficiency in production simply means that given scarce resources shall be employed in such a way that the value of net output (in more technical terms, the willingness to pay for output less the opportunity costs of the inputs) will be at a maximum. From this general statement, efficiency rules may be derived, such as (1) output price must equal marginal costs, and (2) input price must equal the value of the marginal product. Both spatial and temporal dimensions of efficiency exist. While a definitive statement of efficiency would require rigor beyond the scope of this paper, the notions are intuitive enough to be generally understood.[4]

Distributional equity (fairness) refers to the distribution of net output among the agents of society that have claims to it.[5] What these claims are and how they came to be what they are is a complex and difficult subject which cannot be fully discussed here. Still, it is quite apparent that the issue of basic fairness must be raised since so many people see the question of ownership of the public lands primarily in these terms.

This chapter is not about the administration's asset management program, although much of it will be applicable to that program also. Neither is it a definitive case for divestiture of all the federal lands. I regard it as primarily an argument for disposing of the so-called commodity lands, those that produce timber and livestock forage primarily. Of course, other valuable products are produced on these lands as well, so there is no escape from the multiple-use problem.

Current Allocation and Investment Policies on the Public Lands

Perhaps the arguments about retention or disposal of the public lands can best be understood and evaluated by putting them into a framework that consists of three quite general hypotheses: (1) the size and composition of the bundle of products (outputs, including amenity services) taken from the public lands is now suboptimal (the production problem); (2) the quantity and productivity of the resources utilized in generating public land outputs is suboptimal (the investment problem); and (3) the rationing devices used as a substitute for price allocation to distribute the public land outputs among recipient agents is suboptimal (the distribution problem). If the hypotheses are confirmed by relevant argument and data, resources will be inefficiently allocated and wealth will be sacrificed, or income and wealth will be maldistributed, or both.

Testing the hypotheses will not be an easy task for many reasons. What criteria will be employed in the test? If resource allocation is judged by economic efficiency norms, as has been suggested, then economic theory has furnished some simple rules that permit us to differentiate between inefficient and efficient outcomes. It is not very helpful to reject this framework without having another as an acceptable substitute. Without norms, no fruitful evaluation and comparison can occur. My position on this issue is that even though other goals besides efficiency may exist, efficiency problems that are at the heart of public land management can best be analyzed and evaluated by efficiency norms, and thus I do not hesitate to use them. We must be clear, however, that an efficient outcome may not be socially optimal in all respects.

Unfortunately, we have no equivalent equity or distributive norms. Shifts in the distribution of income and wealth resulting from policy changes can be described, but inferences about the desirability of these shifts are impossible to draw without the required norms. Often analysts make assumptions that egalitarian distributions are better than highly concentrated ones, but these assumptions are simply value judgments about which science has little to contribute. Recent literature has developed some concepts of "fair" allocations that appear promising,[6] but so far nothing has emerged that would allow us to be confident enough to say anything very definitive about alternative income distributions.

This conclusion may be too negative because it misses the interrelationships between equity and efficiency. Policies that capriciously deprive some agents of income and wealth in order to transfer them to others in the name of equity might create an environment that militates against efficient production and investment. If so, equity considerations become part of the efficiency problem.

Another important issue surrounding the use of efficiency norms is the problem of unpriced products, amenity values, and resources. I shall discuss some of these problems in later sections.

The evidence needed for testing the hypotheses about efficient resource allocation is of two types: (1) analysis of institutions making allocation decisions to determine if incentives and information are adequate to *expect* efficient decisions; and (2) empirical studies of *actual* decisions that can be judged as efficient or inefficient by selected criteria.

The nature of decisions in public land management

Thomas Sowell has written, "An economic system is a system for the production and distribution of goods and services. But what is crucial for understanding the way it functions is that it is a system for *rationing* goods and services that are inadequate to supply all that people want. This is true of any economic system, whether it is called capitalism, socialism, feudalism, or by any other name."[7] In comparing production and distribution decisions under federal land retention and disposal, two organizational ways of making allocation decisions are being compared, each with its own special characteristics and merits and demerits.

As a general proposition we can characterize decisions under private ownership as market-oriented decisions, whereas those of continuing federal ownership and management are primarily administrative and political. The existence of complex planning units in the federal agencies with their sophisticated optimization techniques and elaborate resource classification schemes does not alter the fact that decisions are fundamentally discretionary and political. If the federal lands were in the private sector, political considerations would not be absent since private resource use is often constrained (regulated) by government. Still, at bottom, private owners will be very sensitive to the amount of wealth they own, and the allocation criteria they use will generally be consistent with maximization of their wealth.

Given the fact of scarcity and the need to ration, the allocation problem is greatly affected by the method of denial utilized in regulating access to goods and services and natural resources.[8] It seems obvious that private entrepreneurs use *price* as the mechanism of denial to the extent that they can. Otherwise, they must sacrifice wealth. Factors that prevent even private owners from so doing will be considered later. It is equally obvious that public land managers do not use price to deny access for the most part, one of the reasons being that the allocation processes utilized by the agencies do not permit the informal bargaining needed to establish market prices.

This is not to say that public managers are not subject to pressures that attempt to influence their decisions. Clearly they are. But in a market, price is what the buyer gives up to acquire a unit of the good or service. Price is also the compensation received by the seller. In the public decision arena, the potential recipients of the benefits from a decision may pressure, cajole, and otherwise attempt to affect the decision. They expend resources in the process. In the absence of bribery, however, the public manager receives no monetary payoff that resembles a price. He may receive approbation, goodwill, and support for his continuing employment from those on whom he bestows his favors. But will he receive monetary benefits equal to the "value" of his decision? I see no *a priori* reason to believe he would, and I have seen no evidence to the contrary.

The implications for economic efficiency, as defined above, are significant. How can a competitive price be equal to marginal opportunity cost if price is not the denial method utilized—indeed, especially if a price does not exist? Even on the face of it, the efficiency norm could be expected to be reached only fortuitously.

Free and competitive prices convey *information* essential to efficient allocation. If consumption of goods and services is only denied by price, then price represents the marginal valuation of all agents consuming the good or service. If the market is in competitive equilibrium, price also represents the marginal opportunity cost. How can allocation be efficient with price equal to marginal cost without this essential information?

One answer that has been given is that in a representative democracy the purpose of government is to advance the public interest, however defined. In a sense, political markets exist with "implicit" prices in the form of votes, lobbying, campaign contributions, and so forth. How do we know that this political market is not economically efficient in inducing the production of the efficient bundle of products from the public lands, stimulating the efficient amount of investment, and effecting an efficient distribution, assuming that one could be defined? The answer to this question consists of several points.

Governments *per se* do not make decisions, people employed by the government do. As the public choice theorists have taught us, agency people are like all the rest of us and can be expected to make decisions consistent with their self-interest.[9] This does not rule out altruism if being altruistic brings pleasure or adds to one's feeling of worth.[10] It is almost tautological, however, to argue that the success of an individual employed in a bureaucracy is inextricably linked to the success of the bureau itself, defined in terms of bureau size, budgets, power, and influence. Contrary to what occurs in a private firm, a political decision maker is seldom in a position to gain personally from reducing agency cost or by selling a product to those who value it most highly, both essential to economic efficiency.[11] The incentive structure in government decisions is simply not compatible with efficiency norms.

It is useful to think of agency decisions about the public lands as a "commons" that is accessible to all, but access is proportional to influence and power. Those who are allocated products at subsidized prices or no price at all tend to be relatively few in number and are generally located conveniently to the public lands. Since what they get is worth more than they pay directly, their economic surplus may be significant. They find it in their interest to do what is necessary to keep the surplus as large as possible. We observe them mobilizing into special interest groups, investing in lobbying, political campaign contributions, and public propaganda in order to increase the probability of decisions made in their favor. The rhetoric predicts dire consequences for society if decisions go against them. The nation's interests become synonymous with their interests, or so they claim. It is commonly observed that if an agency official holds out against these interests, sufficient power exists to see that he will be replaced by another who will be more cooperative.[12]

One reason this process is so inefficient—as well as inequitable, I might add—is that groups competing for products and resources sometimes see themselves as antagonists whose uses are incompatible and mutually exclusive. This has two significant consequences: (1) pressure is exerted for decisions that tend toward single rather than multiple use that may be closer to optimal and are mandated by statute; and (2) as happens in other commons situations, the competition for capture of the allocation decision results in a "beggar-thy-neighbor" strategy, where investment in influence by all parties increases to the point where much of the economic surplus is lost. This is a dead-weight social loss, since the resources expended in the struggle for agency capture could have been utilized for alternative beneficial purposes.[13]

In other circumstances clientele groups may conspire to work together to influence policy in their mutual favor but against the "public" interest.

At the other end of the public land-management spectrum are those providing the bulk of the resources for public management, the taxpayers. Because user fees are seldom set at competitive levels and often are zero, management costs for recreation, forestry, and grazing, for example, are much higher than revenues, implying taxpayer subsidies (see chapter 12, page 280). Then why do the taxpayers not do something about it? Because taxpayers as a group are highly diffuse, most are located far away from the public lands, and as individuals they have a comparatively minor interest in the public lands. Given that their costs of becoming informed about these complex problems are immensely higher than the small benefits captured individually, they remain rationally ignorant and uninvolved. This makes it easy for the concentrated special interests to have their way, and the subsidies continue.

The legislation of recent years, as implemented by the public land agencies, does not require efficient management, despite the interpretation of some observers to the contrary.[14] The 1974 Renewable Resources Planning Act, the 1976 National Forest Management Act, and especially the 1976 Federal Land Policy and Management Act may stress the importance of scientific management, comparing costs and benefits from investment in resources, and other practices that seem to be directing the agencies toward efficiency objectives. In reality, however, the supreme management goals also repeatedly stressed in the legislation are multiple use and sustained yield with public input. The public input issue has already been discussed. By definition, the only groups heard are those having sufficient incentive and resources to influence the outcome of a decision.

If multiple use were construed as producing that combination of multiple products that maximized the net aggregate joint value of these products through time (assuming the tools were available to value all products), the concept would have an efficiency ring to it. But, without prices, how can these valuations be made? What costs should be netted out in estimating net value? Both valuations of outputs and costs present tremendous difficulties for an agency managing the public lands. In practice, multiple-use is simply a recognition that several classes of users have a valid claim to the public land, regardless of the economic efficiency merits. It is the stamp of approval for political allocations that do not force the public managers to face the hard efficiency decisions. Perhaps that is why miners, energy producers, timber harvesters, and graziers love the concept. It is their license to use a parcel of public land, whether or not such use is economically efficient from a broader social point of view. In fact, the Federal Land Policy and Management Act (Section 103) seems to specifically reject economic efficiency norms by defining multiple use as "the management of the public lands and their various resource values so that they are utilized in the combination that will best meet the present and future needs of the American people; . . . with consideration being given to the relative value of the resources and not necessarily to the combination of uses that will give the greatest economic return or the greatest unit output."[15]

Sustained yield is equally amorphous and lacking of economic content. It is usually taken to mean that the resources should be managed so that the level of biological yield of the resources is constant through time. It would be easy to provide examples showing such a policy to be economically inefficient.[16] Efficient economic yields through time will vary in physical quantities offered, depending on rates of the physical yield of the biological resources through time, expected prices over time, expected real interest rates, taxes, and other factors. In some cases it may be economically efficient to use up the entire biological stock and then grow another, a policy that is anathema to the philosophy of sustained yield.[17]

There is still another problem, namely, sustained yield at what level of output? This issue is the cause of much of the conflict between the public agencies and the ranchers over the quantity of allowable grazing. In the field, most public land managers today have been trained in biology in the universities. They see optimal vegetation stands in terms of climax, a very high level of production of usually native plants. Maintaining this range condition at sustainable levels may permit only a small quantity of plant harvest. On the other hand, the rancher is interested in animal maintenance or pounds of beef over some time horizon, implying that his optimum is a much higher level of sustained-yield plant harvest.

Given the forgoing, is it any wonder that Dr. Hagenstein argues in chapter 4 that the current yield of the public lands is very different from what it would be under private ownership and management?

Before leaving this general issue, however, one point should be made on the other side. Informal market transactions do not permit the interests of nonnegotiating parties to be included. If it were costless to bring third parties into the negotiations, they would be included, and no problem would exist.[18] Unfortunately, the transaction costs are often prohibitive. As Sowell notes, "Political systems allow third parties to influence economic transactions from which their interests are excluded. Political decision making can lower transaction costs by allowing relatively few surrogates to make and implement decisions reflecting the will of millions who have [an] insufficient stake (or resources) to incur the huge costs of devising and transacting some of the decisions they believe in."[19] Of course, how significant the external interests are and how well the surrogate (the politicians and bureaucrats) represents them in the political decisions are empirical issues. But, in principle, it must be granted that private decisions will not take these spillover effects into account and political decisions just might. This issue will come up again in a later section.

Thus far, this discussion has been directed to why agency management of the public lands is not likely to be efficient in solving the production and distribution problems. What about the investment problem? Is it likely that under public management the economically efficient quantities of investment funds will be available for use on the public lands? Efficiency would require that investment in improving resource productivity be established at that level where the marginal value of the increased productivity is equal to the marginal opportunity costs of the funds (their best yield elsewhere in the economy). Several reasons exist, *a priori*, for believing that efficient investment will not occur under public ownership.

The discretionary investment resources available to the public land agencies originate from two sources—congressional appropriations and user fees. Since economic efficiency criteria are seldom, if ever, utilized by Congress to appropriate investment funds, we cannot be confident that efficient quantities are appropriated. Political log-rolling and pork-barreling by their very nature are political and not economic calculations.

It is true that recent legislation, such as the Public Rangelands Improvement Act and the Federal Land Policy and Management Act, specifically provide funds for range improvement, some from appropriations to be used at the discretion of the secretary of the interior, and some from designated uses of user fees. An adequate description of the planning processes of the agencies, wherein they decide where and how to invest discretionary funds in resource improvement, is too broad a task to be attempted here.[20] My view is that these processes have been improved in recent years but they are not based enough on economic efficiency criteria.

The user-fee component is allocated by formula to the district or region where the fees are generated, probably as a sop to local users who complained when fees were raised. Investment by formula, however, would always be inimical to efficient investment policy, since funds are funneled to specific purposes and geographic locations regardless of comparative economic yields.

I see two additional problems with current practices of handling user fees. First, they are very low relative to the value of the products. In the case of timber and some energy fuels, competitive auctions are utilized to determine access, and questions have been raised about how competitive those markets are.[21] Grazing is priced below competitive value, as witnessed by the entitlement permit or associated base property taking on value. Other benefit recipients such as water users and recreationists directly pay nothing at all. The upshot of this is that federal agencies receive revenues lower than what a competitive market would produce.

Furthermore, by statute the federal government must share most of these revenues with state and local governments, ostensibly in lieu of paying taxes. This weakens the incentive for the federal government to adopt a fee policy that would price products at value or at supply costs and thus be economically efficient.

Examples of inefficient public land management

Several types of studies of public land management have been made, some relating to the production problem, some to the investment problem, and some to the distribution problem.

In attempting to make the case that private management of the public lands resulting from private ownership would be more efficient than public ownership and management, it would be ideal if actual performances could be evaluated in situations that closely parallel each other. Unfortunately, few such studies have been made. Perhaps the most important reason is that land resources and the products taken therefrom vary tremendously within and among regions and among types of ownership. About the best that can be done is to appraise the efficiency of public management decisions using the efficiency paradigm developed earlier, and infer the changes in decisions that likely would be made if the land were privately owned and managed.

A key study, which looks at both private and public timber harvest policies, is that by Bruner and Hagenstein.[22] They constructed an economic model of the timber industry in the Pacific Northwest. Their findings suggest that if current forest policies continue, the region faces a downturn in harvest over the next two to three decades with concomitant increases in the real price of logs and standing timber. Timber inventories on private lands have been depleted to the point that harvest levels of recent years cannot be sustained. Harvests from the national forests could offset this private decline if policies were changed to increase the allowable cut. It seems clear that if the forests were privately owned, the rate of harvest would increase. It is not obvious from the evidence presented, however, that an increase in the rate of public land harvest would necessarily be economically efficient in the long run. But these results, when combined with those of other studies (to be discussed shortly), do imply "underharvesting" of the national forests.

In the preface to their book, Baden and Stroup say, "When the authority for decision making is not closely tied to the responsibility for outcomes, decision makers have few incentives to consider the full social costs of their actions. The articles in the book strongly suggest that this tendency generates problems in the public as well as the private sector. Thus, we have significant problems in the public management of timber, range, water, and energy resources."[23]

The principal theme of their book is that reductions in environmental quality have been *generated* by the positive action of government agencies.[24] Each chapter speaks to the general theme, but only those reviewed here are specifically applicable to the public lands.

Libecap and Johnson argue that overgrazing on the Indian reservations is a classic commons problem,[25] since it is the result of a failure by the Department of the Interior and the tribal councils to enforce property rights and to adopt an enforceable grazing program. They show further that the problem of overgrazing is exacerbated by the policy of breaking up large herds into small ones, thereby raising the costs of negotiating, transferring, and enforcing property rights. Apparently it was a preoccupation with equity that resulted in the progressive and coerced shrinking of individual herd size and severe restrictions on fencing that caused much of the overgrazing. There can be little question but that privatization of these lands would correct these problems.

Sabine Kremp shows that the BLM has developed a range-improvement practice, called rest-rotation grazing, that is being applied almost everywhere the bureau is active in range improvement.[26] The practice is very costly compared with most other improvement practices and has never been proved to be really cost-effective. Kremp indicates why the bureau benefits from employing such a scheme and cites many range experts who believe other practices are more cost-effective on much of the government land.

The favorite practice of the Forest Service for range improvement is "chaining," reports Ronald Lanner.[27] Chaining is accomplished by two Caterpillar tractors pulling a huge chain between them in order to tear out large vegetation, such as piñon and juniper trees, that compete with grass. Lanner shows the effect on livestock forage (positive), watershed yields (uncertain), wildlife habitat (mostly negative), archeological antiquities (decidedly negative), and pine nuts, a staple of the Indian diet (negative). He reports that Forest Service reseachers claim that only their more successful projects just about break even from a benefit–cost standpoint.

Barney Dowdle and William Hyde evaluate the government's timber management policies.[28] Dowdle argues that a concept called the "allowable-cut effect" is used by agency foresters to justify large budget increases to grow more timber. The federal government owns very large stands of mature timber. The public foresters contend that if they have funds for cultural practices such as planting, thinning, or fertilizing to get more new timber, then by the allowable-cut effect they can justify harvesting more old-growth timber. Dowdle shows how rates of return on this investment in cultural practices have been grossly exaggerated and that, in fact, much of the investment will not yield a competitive rate of return. He also condemns the practice called "cross-subsidizing," wherein timber sales are tied in with the construction of roads. He alleges that by this process, the Forest Service is able to circumvent the budget process and to spend existing timber wealth to implement uneconomic, sustained-yield, evenflow activities.

Hyde shows why it is that free timber markets would more closely resemble the socially efficient allocation of resources than do current harvesting practices of the Forest Service in Colorado. He concludes, "Specifically, disregard for the free timber market leads to a misallocation of public funds in favor of expanding timber management activities to less productive land. This misallocation is reflected in forgone nonmarket environmental and recreational values."[29] The reasons for this are (1) faulty land classification, which overstates the value of timber and results in overinvestment in land for timber production; (2) the definition of maturity implies an excessive investment in capital, or older and larger timber stands; and (3) desired annual even-flows of timber (following sustained-yield imperatives) result in underutilization of mature stands. Hyde's study of the San Juan National Forest in southwestern Colorado empirically substantiates these allegations of misallocation.

In another study of the Douglas fir region of the Pacific Northwest, Hyde also finds evidence of misallocation on the public timberlands: "Profitable output is 60 percent greater than current output and on 17 percent fewer acres. . . . The currently inefficient 27 percent is a sufficiently large portion that it cannot be explained as occasional acres surrounded by better land. It suggests considerable inefficiency in current public land management."[30]

Marion Clawson has also been a long-time critic of the timber-harvesting decisions of the Forest Service. In 1976 Clawson estimated that for the National Forest System alone, inefficiencies in management resulted in forgone annual timber harvests worth $600 million.[31]

The point was made earlier that constraints on management imposed by multiple-use strictures are expected to result in inefficient resource allocation. A 1978 Forest Service study supports the conclusion that multiple-use managerial constraints are serious enough so as to "dissipate all opportunity for timber managerial discretion and optimization."[32]

Most of these studies corroborate the existence of a production and an investment problem as defined above. So far, however, we have not established empirically that there is a distribution problem; that is, the output is not being efficiently allocated to ultimate users.

Consider the regulations to allocate grazing by the Bureau of Land Management. Forest Service procedures are very similar. Grazing Regulation No. 4110.1 specifies that "to qualify for grazing use on the public lands an applicant must be in the livestock business [and] must own or control land or water base property."[33] The base property or "commensurability" prerequisite, as interpreted, requires a rancher to have sufficient fee-simple base property to supply feed and water to livestock during the period they are off the public lands. Thus, the commensurability requirement constitutes a denial for eligibility to those stockmen who do not own sufficient base property, but who may be very efficient in every respect. It is common for ranchers and farmers, especially new ones trying to get established, to rent all or some of their land. Many would not qualify for grazing permits. Use of the public range by certain ranches prior to government reservation and management also has been a significant qualification for permit eligibility. Because the user fee was established below the value of the forage, price could not be the rationing vehicle. Commensurability and prior use were substituted as an equity measure to give preference to local ranchers who had formerly grazed the range.

My own research has demonstrated that this policy has been inefficient in allocating forage to ranchers who valued it most.[34] No doubt this example could be multiplied, as we seldom observe price denial being applied to users of the public lands. The first-come, first-served methods of allocating nearly all recreation services on the public lands are bound to be allocatively inefficient and may reduce the quality of the recreational experience as resources are overutilized.

In sum, there is evidence aplenty to reject the hypothesis that federal management is economically efficient. Theoretical reasoning and empirical analysis both support this conclusion. But could private owners do any better? I turn to that issue next.

Divestiture to Private Ownership

The essential notion is to grant private property rights to own and control the land itself, fee simple, and thus internalize the significant external benefits and costs that under government ownership are a commons, resulting in inefficient production, investment, and distribution decisions.[35] This is a large order since private ownership does not necessarily mean private decisions unfettered by governmental regulations. Land use is often tightly controlled, and we must be clear here that privatization means putting into private hands the fundamental control of land use, parcel size, land transfers, harvest decisions, productivity improvement, and so forth. What would be the result of this? Since divestiture is prospective, very little empirical evidence is available. We can logically infer some results, however.

The benefits from private ownership

Private ownership of the federal lands would create incentives for allocative efficiency in production and distribution for those outputs that can be traded in markets. Equity issues would arise in effectuating the divesting process since those now using the public lands on heavily subsidized terms may expect their income and wealth to be affected. Unless transaction costs are prohibitive, privatization would imply that denial would be accomplished by price. This does not mean, however, that land users would necessarily be worse off, as I shall show later. In any case, once this initial redistribution effect has worked itself out, equity ceases to be an issue with market-traded goods, since presumably no free-market exchanges occur unless both buyer and seller believe the trade will make them better off.

Under private ownership the multiple-use problem does not exist, except for the issue of public goods. Pure public goods are those where consumption is not exclusive and where control over access is difficult and prohibitively costly.[36] When these conditions obtain, price cannot be used as the rationing vehicle and the market will not provide for optimum supplies. Investment resources cannot be acquired and "free-riding" by consumers is observed, since nonpayers cannot be excluded from consumption. The question is, How prevalent are pure public goods on the lands under discussion? Opinions vary widely, but in my view, public goods do exist and are of some importance. I will return to this issue later.

For market goods, owners will maximize their wealth in the resources by matching their production with the preferences for various products by consumers. Price is the coordinating vehicle in providing information. Optimal product diversity would be forthcoming as producers respond to the purchasing power of all consumers who demand different products at different prices. Little doubt exists that products would be more diverse, prices (fees) would be more variable, and many more consumers would be served under private ownership than currently are under public ownership. Consumers and taxpayers would have better and more accessible information and greater incentives to acquire it, and their current rational ignorance would be displaced by an active knowledge of the available products and their prices.

Management decisions would be fully accountable and constrained by the need to cover costs in the long run. As a result, per-unit costs of output would be expected to fall, not only in comparison to public management, but also through time, since incentives would exist to employ cost-reducing technologies. Prices would serve the valuable function of information feedback between managers, consumers, and owners of factors of production.

Resource productivity could be expected to increase, and enhanced conservation would occur. "Underinvestment" exists now because public agencies cannot obtain adequate funding and because private funding is either prohibited or constrained by excessive risk. Privatization would soon correct this situation with the private capital market supplying the resources for all supramarginal investments. Significantly, the risks associated with current political management would disappear. Various types of agreements, covenants, easements, and other instruments would clearly define property rights and reduce risk.

Baden and Stroup have demonstrated why private management is superior to public management in conserving resources.[37] Part of the reason is that elected officials establishing guidelines do not see much beyond the next election. Furthermore, future generations are not here to vote. Bureaucrats are whiplashed by political forces that have great urgency in the present.[38] By contrast, privately owned resources will tend to be owned and controlled by those who are most optimistic about the future. Present wealth in land and renewable resources is the market's expected value of the discounted flow of valuable future products. Optimists, who see higher future values than pessimists do, bid away resources. Future generations *are* represented by entrepreneurs who profit from conserving resources for their expected use. Just as the market ensures efficient allocation to those current consumers with greatest effective demand, so does it ensure optimal allocation to those time periods with expected greatest effective demand. Of course, expectations about the future vary widely and the optimists may be wrong. But that is just the point! If conservation is transforming current assets into future assets, what more reliable and more efficient mechanism for conservation do we have available than the profit motive of these optimistic entrepreneurs operating in the private sector?

Can an efficient disposal program be implemented?

Perhaps this talk of public land disposal is academic. Even if public mismanagement is accepted, can the political surrogates of the American voters be persuaded that privatization is in the public interest? Can disposal schemes be found that will be efficient and reasonably free from fraud and political favoritism, a desirable condition for disposal?

Some proposals have linked public land disposition to reducing the national debt by trading land for extant government bonds.[39] It is believed that the distortions in land and capital markets would be less than if payment were in dollars. A lengthy discussion of this issue is beyond the scope of this chapter. Suffice it to say that I believe the impacts on the land and capital markets would not be great because the scale of divestiture is likely to be small and the pace through time rather slow. Thus, I do not see land purchases being regarded as a substitute for investment in new capital projects in the economy. The Federal Reserve Bank could always alter monetary policy to induce the desired rate of monetary growth.

It has been suggested that putting public land up for sale would depress land prices. I am not so sure. Two effects must be considered. On the one hand, the supply of land to private markets would be effectively increased. Public land to be disposed of would be considered a substitute for private land, and this would have the partial effect of depressing land prices. On the other hand, if my arguments are correct that now public lands would be much more productive under private ownership, then the productivity of complementary private land could be expected to increase. In other words, economic rents earned by land could be expected to increase. This partial effect would be to increase land prices. The net effect is an empirical question and I would not even hazard a guess as to the final outcome. The point is that predictions of significant decreases in private land prices are probably exaggerated.

Actual disposal methods are of much more concern from an equity than from an efficiency point of view. The purchase price will naturally be a function of the specification of property rights, the parcel size, standing growth of biological material, productivity of soil resources, terrain, subsurface mineral and energy fuel wealth, and competition in the land market. But once the sale is consummated, more efficient resource shifts will occur if there are no impediments to transfer and property rights are not severely attenuated.

But equity problems abound. What is to be done about the interests of current users tied up in grazing permit values, improvements privately financed on the current public lands, existing leases that have not expired, and patent claims for minerals. The problem is that it is these very equity problems that will make any disposition politically untenable unless they are resolved.[40]

Perhaps the most serious issue of all is the potential for fraud and special and discriminatory treatment. If political allocation cannot be efficient, why should we believe that political disposal will be either efficient or equitable? The government has a long history of fraudulent and inequitable distributions of land and water. At least this is the public's perception. This is a legacy that will undermine any efforts to convince the public that it will be different this time.

Still, we should have learned a few things from previous experience. Our best chance would result from broad public debate of a disposal philosophy and the details of a disposal system. But I for one am not optimistic that once a suitable policy were developed, it could be implemented without great political turmoil.

An Assessment of the Criticisms of Divestiture

One of the most obvious objections to private ownership and market allocation of products from the public lands is that market goods will be priced at competitive levels rather than received free or at highly subsidized prices as they are now under public ownership. While the point is valid in a general sense, the situation is not so simple as appears at first blush. It would appear that gainers and losers are easy to identify: (1) private owner producers would presumably gain and the local economy would pick up from the extra income generated from better management, as well as from the pricing policy; (2) present bureaucrats would lose and presumably would have to change jobs or move elsewhere; (3) local governments would gain or lose depending on whether local taxes were higher than in-lieu user fees are now; (4) national taxpayers would gain by eliminating costly public ownership and management; and (5) consumers of products may lose or gain.

How can consumers gain if market products are to be priced at competitive levels? The reason is the products under public management are not free or as subsidized as they appear to be. Although free-riding is common, someone must pay for those lobbying efforts, campaign contributions, propaganda, and court costs incurred to influence public decisions. Indirectly, the funds probably come largely from dues paid to environmental organizations and commodity associations. This process is inefficient because these funds are not directly associated with consumption and therefore are not prices that regulate consumption at the margin. It is conceivable that privatization and market allocation at a price would be both more efficient and more equitable. Each consumer would pay for what he gets, and free-riding would not exist. Because resources would be more productive and product diversity would increase as products were tailored toward consumer tastes, the value of outputs to consumers would increase. Even if they paid a competitive price it is conceivable, even likely in my opinion, that consumer surplus would increase.

Thus far I have dodged the problems of public goods and externalities, except to define them. There appear to be some problems with divestiture if public goods cannot be marketed and externalities are not accounted for in negotiated private decisions.

Examples of public goods are open-space amenities and "existence" values. The latter require a brief explanation. Knowing that a wilderness exists may have value to some people, even if they never visit a wilderness area. But why is this value dependent on *public* ownership of the wilderness? A wilderness of comparable characteristics that was privately owned would have no less existence value. Even if they are private, many facilities, such as the Empire State Building and Dodger Stadium, have existence values for some people. So would Death Valley if it were in private hands.

But would there be any wilderness if the land were privately owned? If there is demand for a wilderness experience and access can be controlled, a private owner will supply the experience if costs can be covered with revenues. If demand is not adequate to cover costs, then how can society be better off by having wilderness supplied, apart from the "existence" value? It is the access problem that is critical.

Several amenities such as hiking, backpacking, and much instream water recreation would be difficult to market because access is costly to control. To some extent also, consumption is nonexclusive. The private market will not efficiently supply these amenities, although private lands currently supply large quantities of these public goods because it is simply too costly to forbid consumption. It must be granted, however, that the supply of public goods will be below some theoretical optimum under private ownership. But where is the evidence that their supply is closer to optimum under public ownership? Much of the argument presented earlier would lead to the *a priori* expectation of suboptimality. Only empirical analyses can reveal whether public or private management would be more efficient in yielding public goods, and these studies have not been made. Even so, where it is obvious that losses of public goods would be substantial under private land ownership, it may be possible to regulate the landowner to attain a more efficient mix of public and private goods. Property rights could be attenuated so as to guarantee production of public goods, and this would be clearly established before the land was disposed of to private owners.

Opponents of privatization have made much of the fact that some goods, especially amenity services, are not market-priced and thus cannot be efficiently market-allocated.[41] Therefore, "collective" decisions and, by implication, government ownership and management, are required. Several responses can be made.

Because price denial is not the rationing system utilized for allocating products of the public lands, naturally market prices for these products do not exist. But this fact is irrelevant to the issue under discussion. The right question is, Could and would they be priced and market-allocated under private ownership? The correct answer is yes, at least for all nonpublic goods, which include most of those not now priced. This would include all recreation where access can be controlled. Already we have a wide variety of private markets in hunting, fishing, camping, skiing, boating, swimming, and so forth. There is no reason, at least in principle, why the market could not efficiently allocate these amenities from the current public lands if they were private. The whole issue is largely a ruse to justify political allocations of the kind now extant.

External effects may be highly significant in some situations. Water constitutes an excellent example. Most of the fresh water used for municipal and industrial purposes, irrigation, and recreation originates from precipitation on the public lands of the West (see chapter 11). The condition of the watershed is critical to runoff rates and water quality. Do not these facts imply that society has an overriding interest in these watersheds and that continued public ownership is mandated?[42] I have my doubts for several reasons.

It is by no means clear that water yields and water quality would be lower under private ownership. I have already argued that timber yield and range condition—both positively related to watershed conditions—well might be superior under private ownership. The water yield from the lands under discussion probably would increase under private management.

Perhaps even more important, mechanisms exist that would internalize much of this externality problem should it get out of hand. Appropriative water rights are separated from the land in any event and are determined by the date of filing. Most of the streams originating on the public lands already are fully appropriated and water rights have been sanctioned by state law. If private landowners disturbed these rights in any significant way, they would be liable under the law and subject to court action. My own guess is that conflicts over water rights might well be far less serious if the public lands were privately owned than they are now under federal ownership. Even the threat of a lawsuit might deter private landowners from socially deleterious actions that do not even faze public decision makers.

What about fugitive resources, such as wildlife, that cannot be circumscribed by private boundaries.[43] Does this mean that we must have a public landlord to internalize the externality? The situation is roughly equivalent to the water problem on the production side, but the consumption side is quite different. Hunting and wildlife harvest are reasonably site-specific, and access is controllable to some degree. Already in many places ranchers have found that the fees they charge hunters for crossing private land in order to hunt on public lands exceed their income from ranching. They defend these fees as compensation for the damage incurred from wildlife grazing on their private lands, which they cannot control. It may be that habitat for wildlife would differ under private than under public ownership. I am not sure it will be inferior. It is conceivable that the thousands of private entrepreneurs who are marketing hunting and fishing privileges might well manage these resources more efficiently in order to produce and retain wildlife in areas under their control.

In attempting to justify continuing public ownership of land, Bromley has argued that a distinction must be made between private and social efficiency.[44] Prices do not reflect social values because of externalities—that is, uncompensated costs and benefits. Other classes of market failure, monopoly and monopsony, and lumpiness of outputs and inputs prevent marginal equalities needed for private market efficiency. Therefore, prices have no normative appeal. Thus, public ownership and management are required for social efficiency.

It is quite true, in principle, that market failure can distort market prices to the extent that private markets will misallocate resources. The distortion will be a positive function of the extent of market failure. I have argued, however, that market failure may be far less significant than what it is alleged to be by the opponents of privatization. However, of greater importance is government's failure to efficiently allocate resources. Externalities and public goods may distort prices in private markets. But under public ownership, no reliable prices or even "shadow" prices are utilized at all for allocating purposes, except possibly for timber and energy auctions. There is no compelling evidence that political allocations are efficient in internalizing externalities or providing public goods. Concepts of efficiency simply are irrelevant for all the reasons earlier presented. So why does it necessarily follow that public management is a prerequisite for social efficiency?

It has also been suggested that corporate bureaucrats in the large firms of the private sector are remarkably like public agency bureaucrats in the public sector and thus there would be few gains from trading one set for another. DeAlessi has convincingly argued that this is not so.[45] Private and public organizations differ in the cost of transferring ownership shares. An individual can change his "ownership" portfolio of public benefits only by moving from one jurisdiction to another. This is far more costly than buying or selling securities, which make up his portfolio of private ownership. Thus, property rights in public organizations may be taken to be nontransferable. The owner's incentive to detect and inhibit undesirable managerial behavior is much weaker in public organizations than in private firms, and it gives government decision makers greater opportunities to increase their own welfare relative to that of the owners.

DeAlessi explores the implications of this theory and deduces many of the problems with public ownership which I have discussed earlier. Motives, incentives, and owner-imposed constraints on management are far different in the private than in the public sector, and the private sector more closely meets the requirements of efficiency norms.

Another variant of this private–public ownership issue is that private ownership is unnecessary since "social" purposes will be achieved by public regulation of private management so that the outcome differs only in a minor way from public ownership and management. I know of no evidence that supports such an assertion. In fact, quite the reverse is the case: the public regulators are apt to be "captured" by the private interests that they are supposed to be regulating. This phenomenon has been widely observed empirically and also says something about the relative strengths of public and private interests, and about motives, incentives, and goals.[46] Even if public *managers* have not been captured by clientele groups, as is argued by Culhane,[47] this does not mean that public *regulators* of private owners will not be.

Finally, objections to private ownership—even if it is efficient in resource allocation—have been raised because the resulting income distribution is unacceptable. These objections have been made without proof of any kind. What bothers people is that price denial rather than other forms discriminates against the poor who cannot come up with the funds to pay the price.[48] Price allocation thus makes the distribution of income and wealth more concentrated.

The point was made earlier that we have no scientific methods of comparing the social desirability of alternative distributions of income and wealth. Moreover, it is well known that the bulk of the users of the public lands are not low-income citizens, especially most recreational users. As a rule, their incomes and discounted future incomes are far higher than the average of all taxpayers. Thus, current public land allocation methods transfer income and wealth away from the poor and toward the nonpoor.

Conclusion

This chapter has demonstrated that political allocations of resources, associated with production of commodities and amenities, investment in resource improvements, and the distribution of outputs among competing agents is *inefficient,* when viewed in the framework of traditional neoclassical economic theory. Most of the inefficiencies inherent in political allocations would be overcome if public ownership and management were replaced by the private sector. Some concerns exist with private market allocation because of market failure in the form of externalities and public goods, but it is argued that they are probably less serious than government failure connected with current public management.

Because of equity difficulties that would be part of any disposal program and a healthy public skepticism about the potential for fraud and giveaway, I am doubtful that disposition on a large scale is politically feasible, but economists are often very poor at assessing political feasibility. Most of the efficiency gains are *prospective* and many of the beneficiaries do not even know who they are. Current users would be threatened by privatization unless the disposal policy were clearly favorable to them. For these reasons, I am doubtful that a significant constituency for disposal presently exists. In my view, we must have more evidence that an efficient and equitable disposal policy is available. In addition, the public must be convinced that such is the case.

Let us begin with a small disposal program and be creative in trying alternative schemes. Perhaps in this way the necessary data can be obtained and a persuasive case made for disposal on a large scale.

Notes

1. See Steve H. Hanke, "The Privatization Debate: An Insider's View," *The Cato Journal* vol. 2, no. 2 (Winter) 1982, p. 685.

2. Ibid., p. 655.

3. John Baden and Richard Stroup, "Saving the Wilderness: A Radical Proposal," *Reason* vol. 13, no. 3 (July) 1981.

4. An excellent technical description of efficiency and fairness can be found in Hal R. Varian, *Microeconomic Analysis* (New York, W. W. Norton, 1978) chap. 7.

5. Equity or "justice" as an ethic has been elaborately discussed in John Rawls, *A Theory of Justice* (Cambridge, Mass., Harvard University Press, 1971).

6. W. J. Baumol, "Applied Fairness Theory and Rationing Policy," *American Economic Review* vol. 72, no. 4 (September) 1982, pp. 639–651.

7. Thomas Sowell, *Knowledge and Decisions* (New York, Basic Books, 1980) p. 45.

8. Ibid.

9. Two of the classic works in this area are W. A. Niskanen, *Bureaucracy and Representative Government* (Chicago, Ill., Aldine–Atherton, 1971); and James Buchanan and

Gordon Tullock, *The Calculus of Consent* (Ann Arbor, University of Michigan Press, 1962).

10. See Gary S. Becker, *Treatise on the Family* (Cambridge, Mass., Harvard University Press, 1982) chap. 8.

11. Louis DeAlessi, "Implications of Property Rights for Government Investment Choices," *American Economic Review* vol. 59, no. 1 (March) 1969, pp. 13–24.

12. Richard L. Stroup and John Baden, "Endowment Areas: A Clearing in the Policy Wilderness," *The Cato Journal* vol. 2, no. 2 (Winter) 1982, p. 700.

13. The question of the various influences on the decisions of public land managers has been penetratingly studied by Paul J. Culhane, *Public Lands Politics: Interest Group Influence on the Forest Service and the Bureau of Land Management* (Baltimore, Md., Johns Hopkins University Press for Resources for the Future, 1981). Culhane differentiates capture and conformity. The capture thesis suggests that public land administrators are "captured" by clientele groups, who then dictate policy. The conformity thesis argues that politics are neutral and that field officers subordinate themselves to their superiors who dictate policy on the basis of traditional administrative principles. Culhane argues that the public lands agencies are not captured but are influenced by the clientele groups.

14. John V. Krutilla and John A. Haigh, "An Integrated Approach to National Forest Management," *Environmental Law* vol. 8, no. 2 (Winter) 1978, p. 413.

15. Federal Land Policy and Management Act of 1976 (43 U.S.C. 1701).

16. See B. Delworth Gardner, "Agriculture as a Competitive Segment of Multiple Use," in *Land and Water Use* (Washington, D.C., American Association for the Advancement of Science, 1963), pp. 99–115.

17. See an excellent paper, John V. Krutilla, Michael D. Bowes, and Paul Sherman, "Water Management for Joint Production of Water and Timber: A Provisional Assessment," Discussion Paper D-103 (Washington, D.C., Resources for the Future, February 1983).

18. Ronald Coase, "The Problem of Social Cost," *The Journal of Law and Economics* (October) 1960, pp. 1–44.

19. Sowell, *Knowledge,* pp. 36–37.

20. These problems are more fully discussed in B. Delworth Gardner, "The Role of Economic Analysis in Public Range Management," Giannini Foundation, University of California, Berkeley, Research Paper No. 626. Paper presented at a symposium organized by the Committee on Developing Strategies for Rangeland Management, National Research Council (Washington, D.C., 1981).

21. Ronald N. Johnson, "Oral Auction Versus Sealed Bids: An Empirical Investigation," *Natural Resources Journal* vol. 19 (April) 1979, pp. 315–335.

22. William E. Bruner and Perry R. Hagenstein, *Alternative Forest Policies for the Pacific Northwest* (Pullman, Forest Policy Project, Washington State University, June 1981).

23. John Baden and Richard L. Stroup, eds., *Bureaucracy vs. Environment: The Environmental Costs of Bureaucratic Governance* (Ann Arbor, University of Michigan Press, 1981) p. i.

24. Ibid., p. 4.

25. Gary D. Libecap and Ronald N. Johnson, "The Navajo and Too Many Sheep: Overgrazing on the Reservation," in *Bureaucracy vs. Environment*, pp. 87–107.

26. Sabine Kremp, "A Perspective on BLM Grazing Policy," in *Bureaucracy vs. Environment*, pp. 124–153.

27. Ronald M. Lanner, "Chained to the Bottom," in *Bureaucracy vs. Environment*, pp. 154–169.

28. Barney Dowdle, "An Institutional Dinosaur with an Ace: Or How To Piddle Away Public Timber Wealth and Foul the Environment in the Process," *Bureaucracy vs. En-*

vironment, pp. 170–185; and William F. Hyde, "Compounding Clear-cuts: The Social Failures of Public Timber Management in the Rockies," in *Bureaucracy vs. Environment*, pp. 186–202.

29. Ibid., p. 187.

30. William F. Hyde, *Timber Supply, Land Allocation, and Economic Efficiency* (Baltimore, Md., Johns Hopkins University Press for Resources for the Future, 1980) p. 139.

31. Marion Clawson, "The National Forests," *Science* vol. 119, no. 4227, (Feb. 20) 1976, pp. 762–767.

32. Roger D. Fight, K. Norman Johnson, Kent P. Connaughton, and Robert W. Sassaman, "Roadless Area-Intensive Management Tradeoffs on Western National Forests," USDA Forest Service, 1978; and Hyde, *Timber Supply*. p. 41.

33. See Gardner, "The Role of Economic Analysis," p. 13.

34. B. Delworth Gardner, "Transfer Restrictions and Misallocation in Grazing Public Range," *Journal of Farm Economics* vol. 43, no. 1, 1962, pp. 50–63.

35. For an excellent discussion of this issue, see Harold Demsetz, "Toward a Theory of Property Rights," *American Economic Review* vol. 57, no. 2 (May) 1967, pp. 347–359.

36. Paul A. Samuelson, "The Pure Theory of Public Expenditures," *Review of Economics and Statistics* vol. 36, pp. 387–389.

37. John Baden and Richard L. Stroup, "Transgenerational Equity and Natural Resources: Or, Too Bad We Don't have Coal Rangers," in *Bureaucracy and Environment*, pp. 203–216.

38. See Demsetz, "Toward a Theory," p. 355.

39. See Hanke, "The Privatization;" and Gordon Tullock, "The National Domain and the National Debt," Paper presented at RFF Conference on "Rethinking the Federal Lands," Portland, Ore., September 1982.

40. A really creative proposal for disposing of the public lands has been made by Vernon L. Smith, "On Divestiture and the Creation of Property Rights in Public Lands," *The Cato Journal* vol. 2, no. 3 (Winter) 1982, pp. 663–686.

41. Daniel W. Bromley, "Public and Private Interests in the Federal Lands: Toward Conciliation," Paper presented to a conference on "Federal Lands and the U.S. Economy: Striking a Balance in the 1980s and the 1990s," sponsored by the Wilderness Society, Airlie, Va., November 15–16, 1982.

42. Ibid., p. 7.

43. Ibid., p. 6.

44. Ibid.

45. DeAlessi, "Implications."

46. It was principally this contribution of private capture of public regulators that won George Stigler the Nobel Prize for Economics in 1982.

47. Culhane, *Public Lands*.

48. For example, see Bromley, "Public and Private Interests," p. 13.

9

Ownership and Outcome

An Economic Analysis of the Privatization of Land Tenure on Forest and Rangeland

GORDON C. BJORK

It is time to reevaluate the ownership and management of the federal lands. There are two reasons for the timeliness. The first is economic; increases in the relative prices of wood and animal products from the public lands and increased values placed by our society on recreational activity and watershed functions of the federal lands make them much more valuable than they were a half-century ago. The second is intellectual; lawyers, economists, and political scientists have raised public consciousness about the effects of differing distributions of property rights and political arrangements for the uses of and benefits from the public lands.

Shifts in relative factor prices produced by technological and demographic changes have been used by economic historians to explain institutional changes in economic, political, and legal arrangements.[1] The argument is that property rights controlling asset use will be changed when it becomes worthwhile for private groups who would benefit from a privatization of previously scattered, insecure, or commonly held ownership rights to make the investment in institutional change. The argument could apply today to ranchers with BLM grazing permits, wood-products companies relying on Forest Service lands, or the Sierra Club.

This chapter will examine the relationship between the technologies and social values affecting land use and the definition and distribution of property rights controlling land use. I have argued elsewhere that property rights are continuously redefined to take into account changes in technology and social values.[2] In this chapter, I will argue, simply, that changes in forestry and range management practice, changes in wood-product prices, and the imputed values of such noncommodity flows from land as recreation and watershed maintenance make some alterations in the tenure of forest and rangeland rational social choices to increase the net value of output from those lands.

Any existing pattern of property rights is an inheritance from the past. Professors Gates and Dowdle have given alternative explanations of the process of disposal, retention, and reacquisition of public lands in chapter 3. Property-rights theorists explain the historic social assignment of property rights in land in terms of the benefits of "internalizing" the costs and benefits of using land to increase net benefits. To use a simple and timely example, the private owner of a plot of forestland will harvest it when the increase in the value of the trees no longer exceeds the costs of waiting. He will reforest the land if he expects that the present value of a tree harvest in the distant future exceeds the present value of the costs of reforestation. If a person has no property rights, he or she will take the trees now, before someone else does. And the individual certainly will not bear the expense of reforestation if he or she has no enforceable claim to the future benefits. Thus, property rights ensure economic utilization of scarce resources.

The argument for property rights in land is not an argument for private ownership. Public ownership and management could, theoretically, maximize net benefits from land in the same way as private ownership. Historically, private ownership of land in the United States has been the dominant mode of land tenure for reasons relating to Lockean political philosophy and the economic theorizing of Adam Smith and his successors.

The strongest economic efficiency arguments for the private ownership and management rather than public administration of federally owned forest and rangelands have been made in recent years by a group of Adam Smith's intellectual heirs who call themselves "public choice" theorists.[3] They dispute the assumption that benevolent bureaucrats will scientifically and economically manage the public lands in the public interest. They argue that government managers do not seek to maximize the welfare of the nation but, rather, they seek to maximize the income and power of their bureau.[4] Since it is possible, under restrictive neoclassical economic assumptions, to show that maximizing the wealth of the private owners of a resource would also maximize social utility, the conclusion of the public choice theorists is that private ownership would be more efficient than public ownership in attaining social objectives.

One critical element in the argument for public ownership of forest and rangelands has been the interdependence of multiple uses. A forest may provide valuable watershed and recreation benefits as well as harvestable timber but the private owner may not be able to "internalize" or capture those returns and, therefore, does not take them into account in his calculations about the management of the multiple-use forestland. What economists term "externalities" and "public goods" arguments for public ownership, or control, or both, have long been advanced as efficiency arguments for the public lands. The absence of effective property rights to ensure management of multiple-use lands has made public ownership and administrative allocation seem like an "efficient" solution to the management of land.

Public choice economists have questioned the necessity of public ownership for multiple use of forest and rangeland. They point to the current provision of hunting and fishing and even developed camping and picnic areas by private corporate landowners. Watershed stabilization is not a use which could be denied by private owners, and sound forestry management assures its continued provision. They conclude that public ownership is not necessary for the provision of multiple-use management, and they argue that it provides wasteful and inefficient forestry and range management.

Joseph Sax (see chapter 6) does not make the traditional "externalities" or "public goods" arguments for public ownership and management of the federal lands. Professor Sax has made the ultimate argument for government ownership of even ordinary natural resources. While making the lawyer's standard disclaimer that ownership *per se* is of little consequence and that property rights and environmental regulations can be drawn to secure any desired pattern of land use, he proceeds to develop bases for public ownership which I will label the "patriotic" argument and the "demonstration" argument.

Briefly stated, Sax argues that the values placed on privately owned assets by individuals may be less than the value placed on those assets in collective ownership when the same individuals are part of a collective such as a state. Thus, the mere fact of public rather than private ownership may enhance the values derived by the public from the government ownership of lands beyond any flows of benefits from timber, forage, watershed enhancement, or recreation. Linked to this argument is a similar reference to the importance of ecologically sound management as a demonstration effect in forming social values. Professor Sax's patriotic and demonstration arguments for public ownership go beyond any conceptual framework economists have available for evaluation. I will, however, attempt to roughly estimate the magnitude of the op-

portunity costs of potential smaller commodity production in my discussions of the outcomes from different ownership patterns.

Who gets what? The *distributional* arguments for public ownership of
natural resources emphasize the safeguarding of future generations from
rapid depletion of resources.[5] While these arguments are usually made
with reference to nonrenewable resources, they are also applicable when
the private discount rate is higher than the biological yield rate.[6] The
federal lands have been retained, in part, to protect future generations
from this problem.[7]

Distribution of the benefits from renewable resources among the present generation are also an important issue. Should the Intermountain
West be strip-mined to air-condition Chicago? Should Sierra Club environmentalists have their hiking trails preserved at the expense of the
Oregon economy and higher lumber prices for the general public? Economists can make estimates of distributional consequences. Their distributional recommendations, when based on ethical choices, should carry
no more value than Everyman's. Ethics are prior to politics, and the
political assignment of property rights determines distributional consequences which economists may analyze positively but must evaluate
normatively. I will not comment on them further in this chapter.

Biology, Technology, and Economics—
The Changing Role of Forestland

How should forestland be managed? There are six critical factors which
must be considered for economic management of forestland: (1) biological yield functions; (2) costs of reforestation and such intensive silvicultural practices as thinning, limbing, fertilization, disease and pest
control, and fire protection; (3) costs of logging; (4) relative prices of
forest products—dimension lumber of various grades, poles, chips, and
others; (5) the rate of discount used to compare costs and revenues over
long time spans; and (6) the value of noncommodity uses such as recreation and streamflow stabilization which may not be realizable by the
forest owner. The arguments for and against privatization of lands currently in public ownership really depend on these variables.

I wish to argue briefly that these variables have changed, on balance,
in such a way as to increase the desirability of private ownership and
more intensive silviculture on public lands with high potential timber
yields.

Until this century, trees were *not* considered a crop. They were viewed
as a standing inventory to be liquidated as quickly as possible so long

as the price for lumber on fluctuating markets was above the cost of logging, milling, and transporting the commodity to market. The rapid harvesting of the nation's forests, without regard to reforestation, made economic sense as a matter of national policy in the nineteenth century.

Trees grow slowly. The costs of reforestation, compounded until the time of subsequent harvest, could not have been recovered by a forest owner. Forestland, as distinct from its standing inventory of trees, had no value. Under these conditions, private ownership made no sense and some forestlands were never alienated into the private domain while others reverted. All that has now changed for reasons connnected with biology, technology, and economics.

Forests can now be grown faster. Some southern pine plantations have a maximum economic rotation of thirty years. Some stands of Douglas fir can be profitably harvested on a fifty-year rotation. Part of the faster crop rotation is biological. Improved genetic strains and modern silvicultural practices make trees grow faster.[8]

Most of the reasons for intensive silviculture are technological and economic. First and foremost, the market price of lumber and wood products has risen. Over the last forty years the wholesale price index for lumber and wood products has risen tenfold while the GNP implicit price deflator has risen only sixfold. While this could be explained as slower technological change in the forest-products industry than the rest of the economy, evidence on relative productivity growth for subperiods for segments of the wood-products industry refutes this. I infer, therefore, that higher relative prices for timber products and increased productivity in the forest-products industry have combined to substantially increase the value of land suitable for timber growing. The liquidation of much of the standing inventory of the nation's forests has made the market price of timber reflect its cost of production and a "reserve price" which reflects its expected future value. With the passage of time, more timberlands will reflect greater and greater differential Ricardian rents with the relative rise in timber prices.[9]

Intensive timber production and shorter crop rotations have also been made economic by technological change. The relative costs of road-building, maintenance, and reforestation have been decreased by the use of modern power equipment. This decreases costs of logging and reforestation and increases the net return to the forest landowner from the sale of logs. The chain saw, the skidder, and the self-loading logging truck have all decreased the relative costs of logging smaller trees and increased the returns from shorter timber crop rotation.

Changing technology also has had a decisive effect on the changing pattern of forest-products prices. The conversion of wood chips to par-

ticle board and pulp has made the utilization of small trees more economic. Veneering and lamination has reduced the premium for clear
lumber and large timbers for structural purposes. These changes have
tended to reduce the economic value of older, larger trees relative to
the younger, smaller trees and forest management has been moved
toward shorter yield cycles, which tend to maximize sheer yield of cubic
feet of fiber rather than dimension lumber of clearer grade.

The influence of the interest rate on timber-growing patterns is
straightforward from the management viewpoint (although it may be
more complex as a matter of theoretical welfare economics in allocating
costs and benefits between generations). A profit-maximizing owner of
forestland will harvest it when its rate of increase in yield falls below
the interest rate. This is why it is economic to harvest old-growth forests
with minimal annual growth if lumber prices are expected to remain
stable over the foreseeable future. The interest carrying cost was the
reason that private owners would not bear the costs of reforestation or
scientific silviculture before the last half-century. The costs of waiting
were greater than the benefits.

One argument for public ownership of forestland is based on the
interest rate. It is argued that the social rate of discount is less than the
market rate of interest paid by the private owner since the latter contains
a premium for risk and uncertainty. The advantage of public ownership
is the lower cost of interest for holding a forest unharvested. Without
going into the social efficiency and distributional aspects of this argument, it is obvious that reducing the length of rotation for forests decreases the relative importance of this argument and weakens this case
for public ownership.

Finally, one comes to the multiple-use argument. Is the value of the
stream of noncommodity uses from forestlands likely to be less under
private ownership and management than it would be under public? One
factor to consider here is whether the value of watershed or recreation
systematically increases with the length of forest rotation. Young forests
may be more subject to fast runoff and erosion and higher transpiration
rates. This would support the argument for public ownership and a
longer crop rotation.

I see no systematic relationship between the age and density of forests
and recreation values. Hikers may like old-growth forests. Deer hunters
may favor new growth. A key element in the recreation value arguments
is whether these uses would be less under private than public ownership.
Many large timber growers provide developed recreation sites to the
public without charge. Similarly, there are probably not substantial differences in recreational access on public and private forests and range-

lands. And if there were, alternative arrangements to public ownership could be made to ensure public access.

To summarize briefly, changes in biology, technology, and economics have increased the economic yields and decreased the economic costs of intensive silviculture. Economic rents, the difference between costs and returns, can be increased by the intensive silviculture assumed more likely to be economically efficient under private than public ownership. Similar analyses can be made for rangeland (although space precludes developing these arguments here). If there are no offsetting losses in the provision of such noncommodity values as recreation, it is arguable that private ownership and management of present public land could be viewed as a timely adjustment of property rights to changes in underlying production functions in forestry.

Land Tenure Implications of Changes in the Economics of Forestry

Professor Sax's chapter concludes with the observation that our nation is presently in transition from a set of property rights drawn to maximize commodity production to a different set designed to preserve things the way they are. "Doing nothing is viewed as the highest and best use of land," he claims. In this section, I will develop the argument, *contra* Sax, that some of the federal lands should be used more intensively for commodity production than they are presently under government ownership. My argument is based on changes in the economics of forest management.

I start from the simple proposition that more is better than less—if private ownership of some of the public lands can produce more wood products at lower cost over a period of time, and there are no offsetting losses of such noncommodity uses as watershed or recreation, then it is in the public interest to privatize ownership.

There is an interesting historical parallel for the present issue of privatization of public lands. In eighteenth- and nineteenth-century England, a series of Parliamentary Acts of Enclosure were passed to extinguish and consolidate multiple-use rights to land. The purpose was to allow intensive agriculture on land formerly grazed or forested with shared rights. The Enclosure Acts were economically advantageous because technological and economic changes raised the potential economic yields from land. But these increased yields could only be realized by a change in property rights which gave private owners incentives for capitalization. The Enclosure Acts were opposed in eighteenth-century England on many of the same philosophic and distributional grounds that the privatization of public lands is opposed today. The increased agricultural output that they made possible was an important contribution to per capita real incomes in a much poorer society. In relative terms, the increase in U.S. per capita incomes made possible by more intensive management is relatively small because the forest products industry is a small part of the national economy. Nevertheless, a change in property rights which reduces costs and increases output deserves serious consideration.

How could we estimate the magnitude of the change in output from more intensive land management? To whom would it accrue? What would be the second-order effects?

First, the difference between the value of output and costs is a measure of "social savings." The social saving is the difference between the value of the output reflected in its market (or imputed) price and the value of the inputs of labor and capital necessary to produce the output. An increase in social saving from more-intensive management of forestland is equivalent to an increase in the gross national product.[10]

Second, the social savings will accrue to the property owner as a surplus over the variable costs of production—an economic rent. They will be capitalized into the value of the land. The capitalized value of the rents will be the theoretical upper-limit price which the land ownership title would command in a perfectly competitive factor market. The maximization of social savings is equivalent to the maximization of economic rents.[11]

Third, any net benefit to the public (as opposed to the private party securing the title to what was previously public land) would occur only because of the adjustment of output to increase economic rents from the land in question. That is a necessary condition. The sufficient condition is that the acquiring party pays a price for the land which allows the public to capture part of the increase in output resulting from the change in property rights. To the extent that the price paid by the private party is less than the capitalized value of the rents, the private party captures the returns.[12] (Actually, since the increased returns are going to be subject to taxation, it is difficult to imagine a situation in which the general public would not gain.)

How great would the gains from more-intensive use of the public lands be? I would "guesstimate" that the value of the federal lands for commodity production (exclusive of the value of mineral rights and standing timber) might be in the $100 billion to $200 billion range. Using a high 10 percent interest rate for estimation, this would, in turn, indicate an annual social saving from their utilization of $10 billion to $20 billion per year. A doubling of social savings from privatization would amount to between 0.5 and 1 percent of the GNP.

My figures are not meant to indicate any degree of precision. They do indicate an upper-limit order of magnitude for efficiency gains from privatization which public choice theorists advocate and an opportunity cost for the nonquantifiable gains from patriotic or demonstration effects, which Professor Sax advocates, on the order of $100 per capita.

Fourth, my discussion has assumed maximization of net present value by private owners: This is the orthodox assumption of neoclassical economic theory. It has, however, been increasingly questioned by organizational theorists. On the one hand, it has been argued that the managers of private corporations will tend to maximize present profits at the expense of the future. On the other hand, it also has been argued that private corporate managers will act like public bureaucrats and increase the volume of production beyond the point of profit maximization to increase the size and power of the corporation which increases managers' compensation.[13] Still another critique suggests that potential supernormal profits (economic rents) will be dissipated in increasing salaries and perquisites for managerial and nonmanagerial employees.[14]

But what are the management objectives of public managers? The public choice school of economists has argued that government bureaus attempt to maximize the total size of their appropriations and their staff.[15] One critic of the BLM's grazing policies has argued that the bureau has done this very effectively over the past two decades. Libecap cites Gifford Pinchot's model for the Forest Service in building up a scientific staff to measure and increase *biological* yield from land as the model used by both agencies to increase employment and increase biomass volume by limiting grazing and cutting.[16] Other critics of the BLM have contended that their expenditures on rangeland improvement far exceed any economic benefits from increased forage which could be realized from them.[17]

Finally, it should be remembered that our discussion of the sale of public land assumes that only a small incremental sale is made. The sale of a substantial quantity of the public lands would qualify the analysis in the following ways: (1) There would be a reduction in the prices of the commodities produced on the public lands (wood products, meat, wool) dependent on the size of the change in supply relative to total supply, and the elasticities of supply and demand for these commodities. Perry Hagenstein, in chapter 4, has estimated that Northwest timber prices would fall 80 percent in the 1980s if federal timber were harvested on the same criteria as private timber. Perhaps this explains the silence on the privatization debate of timber companies who are large land-owners!

Concluding Remarks

Economists make prescriptive conclusions about public policy on the assumption that the maximization of net output, measured in market prices (or imputed market prices) is equivalent to the maximization of social utility. As Joan Robinson, an eminent critical practitioner of neoclassical economics, once remarked, "Utility is a metaphysical concept of impregnable circularity." Much of the theorizing about the appropriate forms of ownership for public lands assumes away the problems public ownership was instituted to remedy.

To make recommendations about social policy, one must first assume that market prices do represent the marginal utilities of all economic flows. Then, one must assume that interpersonal comparisons of utility can be made and measured in cardinal terms. Additionally, it must be assumed that the existing distribution of wealth represents some social objective since that distribution determines allocation and distribution. Finally, it must be assumed that the intergenerational flows of resources resulting from the use of the present value criterion distributes consumption between generations in such a way as to maximize utility.[18] These assumptions underlie my conclusions and the arguments of the privatizing public choice theorists for the superiority of private ownership and market solutions. Belief in "the market" as the ultimate arbiter of public policy involves commitment to a metaphysical system.

Just as a belief in "the market" underlies the public policy choices of many who would increase the privatization of land, a belief in the preservation of nature is the metaphysical basis of the public policy choices of many who would call themselves "environmentalists" and lock up the public lands in a state of nature. Kenneth Boulding, past president of the American Academy of Science, has said of ecology as a basis for public policy

All this suggests that we must look at the world, and indeed at the universe, as a total system of interacting parts. There is no such thing as an "environment" if by this we mean a surrounding system that is independent of what goes on inside it. Particularly, there is no sense at this stage of evolution on earth in talking about "the environment" as if it were nature without the human race. It makes sense to divide the totality of the universe into parts that have some degree of independent dynamic pattern, but none of these parts are really independent of others; all interact. Everything is the environment of something else. When we talk about the environmental problem, we are talking about the total state of the world and evaluating it from the point of human values. We are not talking about the nonhuman part of the system and evaluating it by its own values, because it does not have any.

The evolutionary vision is unfriendly to romantic nature worship, the view that the human race and its artifacts are not part of nature and that nature without the human race is somehow wise and good. This is a fallacy into which many people of goodwill fall. The human race has been produced by the evolutionary process and so have its artifacts. The automobile is just as "natural" as the horse. It is just as much a species, just as much a part of the total ecological system, and the idea that there is something called "ecology" in the absence of the human race and human artifacts at this stage of the development of the planet is romantic illusion. The human race, of course, needs to be aware of its impact on the total system, but the concept of the "environment" as removed from human endeavor is largely illusory. When we talk about the environment, we mean the evaluation of the total state of the planet according to human values, which are the only values we know very much about. If "nature," whatever that is, has values, they are unknown to us. Certainly nature is no respecter of species, and all species are endangered. Nature cares no more about the whooping crane or the blue whale than she did about the dinosaur. Indeed, personification of nature is a romantic substitute for religion without the intellectual substance of a practical argument for faith in the divine as an act of will and commitment.[19]

Professor Sax's argument for the beneficial effects of ecologically sound management of "ordinary" lands makes some judgments about the divergence of human and natural values which are widely held but questionably based.

Where do the ideas of economists and political philosophers leave decision makers? Where do the analyses of the biological, economic, and technological determinants of public and private ownership of the present federal lands leave us? I see them leaving us ready for a consideration of Marion Clawson's incremental proposals for partial privatization by a modified bid process and subsequent empirical comparison of cost and yield results between private and public ownership of similar parcels of land.

Notes

1. D. C. North and R. P. Thomas, *The Rise of the Western World: A New Economic History* (Cambridge, England, Cambridge University Press, 1973); and T. L. Anderson

and P. J. Hill, "The Evolution of Property Rights: A Study of the American West," *Journal of Law and Economics* vol. XVIII, no. 1, (1975) pp. 163–179.

2. Gordon C. Bjork, *Private Enterprise and Public Interest: The Development of American Capitalism* (Englewood Cliffs, N.J., Prentice-Hall, 1970). *Life, Liberty and Property: The Economics and Politics of Land Use Planning and Environmental Controls* (Lexington, Mass., Lexington Books, 1980).

3. Readers unfamiliar with the public choice approach to the analysis of public sector behavior are referred to J. M. Buchanan and R. Tollison, eds., *Theory of Public Choice* (Ann Arbor, University of Michigan Press, 1972); and to the journal *Public Choice* (edited by Gordon Tullock).

4. W. A. Niskanen, *Bureaucracy and Representative Government* (Chicago, Ill., Aldine–Atherton, 1971).

5. Talbot Page, *Conservation and Economic Efficiency* (Baltimore, Md., Johns Hopkins University Press for Resources for the Future, 1977) chap. 7–10.

6. Robert M. Solow, "The Economics of Resources or the Resources of Economics," *American Economic Review* vol. LXIV, no. 2, (1974).

7. Marion Clawson, *The Federal Lands Revisited* (Washington, D.C., Resources for the Future, In press) chap. 2.

8. For empirical estimates of biological yield functions for representative Oregon Douglas fir land under various density and management-intensity assumptions, see J. H. Beuter, K. N. Johnson, an H. L. Scheurman, *Timber for Oregon's Tomorrow* Research Bulletin 19 (Corvallis, Forest Research Laboratory, School of Forestry, Oregon State University, 1976).

9. For price data, see *Economic Report of the President*, 1983, tables B-3, B-59. For productivity trends, see J. Kendrick, *Post-War Productivity Trends in the United States, 1948–1969* (New York, National Bureau for Economic Research, 1973). As an owner of a small wood lot and interested casual observer, I have contrasted tax sales of cutover lands in Oregon in the 1930s at $1 per acre with similar prices for land, *ex* timber, of $200 to $300 per acre in the 1980s.

10. Bjork, *Life, Liberty*, chap. 6.

11. Bjork, *Private Enterprise*, chap. 5.

12. Ibid.

13. William J. Baumol, *Business Behavior, Value, and Growth* (New York, Macmillan, 1959).

14. Nathan Rosenberg, "Some Institutional Aspects of the Wealth of Nations," *Journal of Political Economy* vol. LXVIII, no. 6, (1960) pp. 557–570.

15. Niskanen, *Bureaucracy*.

16. Gary D. Libecap, *Locking Up the Range: Federal Land Controls and Grazing* (Cambridge, Mass., Ballinger Publishing for Pacific Institute for Public Policy Research, 1981).

17. J. B. Stevens and E. B. Godfrey, "Use Rates, Resource Flows, and Efficiency of Public Investment in Range Improvements," *American Journal of Agricultural Economics* vol. LIV, no. 4 (1972).

18. Page, *Conservation*, chap. 9.

19. Kenneth Boulding, *Ecodynamics* (Los Angeles, Calif., Sage Publications, 1978) pp. 19 and 31.

IV
Intermediate Positions and Special Problems

10

Major Alternatives for Future Management of the Federal Lands

MARION CLAWSON

A number of recent events provide evidence of a substantial and growing dissatisfaction with the management of the federal lands over the past decade. In this chapter I will focus solely on the lands managed by the Forest Service and the Bureau of Land Management (BLM).

The workshop at which this material originally was presented, at which academics, professionals, and others representing diverse interests were present, is itself evidence of this dissatisfaction. If everything were perfect or the complaints were minor, there would be little point in expending the time, energy, and funds necessary for a workshop. Yet, in a single year I have participated in three conferences whose objective was to consider the possibilities of substantial changes in federal land administration.

The Subcommittee on Public Lands and Reserved Water of the Senate Committee on Energy and Natural Resources held workshops on this general subject in 1981 and 1982.[1] More formal congressional action resulted in the Resources Planning Act of 1974, the National Forest Management Act of 1976, and the Federal Land Policy and Management Act of 1976, each of which expressed congressional interest and policy guidance for federal land management and reflected, at least to some degree, a dissatisfaction with then-current federal land management. The administration proposals for accelerated leasing on the Outer Continental Shelf (OCS) and its support for privatization of some of the federal landholdings indicate different concerns.

Clearly, the "Sagebrush Rebellion" is an outgrowth of substantial dissatisfaction with existent management that has been voiced by some of the user groups and the elected officials in the states with large acreages of federal lands. Complaints have been made by the commercial users of the federal lands, and the conservation community has expressed its concern over federal land management, especially regarding the establishment of wilderness areas.

These various expressions of dissatisfaction stem from a commonly held concern that the present management practices, to some degree, are unsatisfactory. There is no agreement as to what is unsatisfactory, or as to the kind of changes that should be made. Some degree of dissatisfaction is inevitable, of course, and it may be argued that some or most of it is normal or, at most, transitory. Any significant change in federal land management—whether it is initiated by the executive branch or by Congress—almost certainly will be both preceded and followed by debate and criticism. At this time it is impossible to say exactly what changes, if any, will be made over the next several years. But it is precisely because dissatisfaction has been expresssed and it is possible to effect changes that it is currently desirable to outline the possible alternatives for change. Thus, to a degree, we will be able to inform and stimulate the debate, without prejudging or foreclosing it.

The Changed Role of the Federal Lands

The federal government has owned land, in the proprietary sense, for 200 years. The history of the federal lands has been considered in detail by Paul Gates in chapter 2; what follows here is my brief interpretation of that history.

A federal estate was created on October 29, 1782, when the Confederation Congress accepted the donation of publicly owned lands outside the boundaries of the former British colony, the newly independent state of New York. From that small beginning, the federal estate expanded greatly by treaty, conquest, and purchase, culminating for land areas with the purchase of Alaska in 1867, and with the accession of the Outer Continental Shelf within the last twenty-five years. Under philosophical and legal principles that stressed the desirability of private landownership, about two-thirds of the federal land has been disposed of to individuals, corporations, states, and others. The disposal process dominated federal land history throughout the nineteenth century. This was superceded—at first gradually and later fully—by withdrawal of the federal lands from disposition, first for the forest reserves (now the national forests) and later for the grazing districts. The withdrawn lands have been managed in three fairly distinct eras; by custodial management from the date of withdrawal until about 1950; by intensive management from 1950 to 1970; and by consultation–confrontation management since 1970. For this last period, user groups have been consulted during the planning process in order to minimize the likelihood of future lawsuits originating from a disgruntled special interest. This history is not only long, as judged on the scale of U.S. national existence, it is also complex and, to many of us, intensely interesting.

Why has extensive public ownership of land been sought by many—and at least tolerated by others—in a country which prides itself on the freedom and the virtues of individual action and choice? One reason is that the federal lands have been open to use by private individuals under a wide variety of laws. Individuals chose where and when to enjoy outdoor recreation on the federal lands, ranchers run their livestock on federal lands, forest industry firms cut timber from federal lands, oil, gas, and other minerals exploration is undertaken by private firms, all operating under applicable law and federal regulations. Never has there been any major exploitation of the resources of these lands by federal corporations or agencies. And it is precisely in the interface that exists between public ownership and private use that the difficult problems of federal land management have arisen. If there had been no use, there would have been few, if any, management problems. The intensity of the problem is largely caused by the intensity of the private use and conflicting claims which are made by potential users.

The history of the federal lands shows clearly that change has dominated their use and management over the decades. As the country has grown and developed, the role of the federal lands in the economy, political structure, and society of the nation also has changed. There is no reason to think that all changes have occurred in the past; on the contrary, it is highly probable that major changes will take place in the future, although we cannot tell exactly what they will be. If the history of public lands teaches us anything, it is that history does not stand still. There have been periods of relative quiescence, or of slower change, as well as periods of rapid change, ferment, and uncertainty. Twenty years or more from now we may find that the 1980s have been a period of relatively rapid change.

In retrospect, the 1950s and early 1960s seem a relatively static period, although at the time this was not the observation of federal land managers. I speak not only as a student of the literature, but as one who was a participant. At the time, a number of policy issues seemed to be settled or at least not up for serious challenge. The use of federal lands for virtually all purposes was expanding under applicable laws; accommodations to such increased use were made by the federal agencies, with some conflicts, to be sure, but, in general, more use was fitted into nearly constant acreages. The issue of federal versus private landownership seemed settled in the main, and there was little reason or inclination to argue this issue. The Public Land Law Review Commission (PLLRC) in 1970 advocated that federal land should be disposed of only when there was a clear national interest in such disposal,[2] a view that some had publicly stated earlier.[3] Since 1970, major legislation endorsed the idea of continued federal land ownership on approximately the present scale. Until the last three or four years, there had been no serious debate about extensive further disposals of federal land; that policy issue seemed settled.

But times have changed since those years which, in retrospect, seem so quiet. The demands on the federal lands have increased greatly, and at a level that makes conflicts among users more common, more difficult to resolve, and more emotionally and politically charged. Although it still is possible to accommodate additional uses on the same areas, it becomes more difficult to do so, and it is equally hard to achieve a consensus on management among rival users. In addition, the users have changed: today they are better educated, more widely traveled, more prosperous, and more demanding of their "rights" than were their parents and grandparents. There are many well-trained professionals outside of the federal agencies who are willing to challenge the professional judgments of the agency personnel. Users of public land are no longer willing to accept uncritically the judgments or the pronouncements of federal agency personnel.

On the purely economic front, times also have changed for the federal lands. The capital value of the federal lands has risen dramatically in recent years. No comprehensive, thorough appraisal of those lands has ever been made, but on the basis of a simplistic capitalization of income flows from these lands, I would estimate that the value of the lands and resources administered by the Forest Service and BLM today approximates $500 billion and that in real (constant dollar) terms that value has risen by twenty-two times since the mid-1920s.[4] Such an estimate is crude at best, and undue weight should not be placed on the precise figure for the current estimate or on that of the estimated increase. No informed observer will question that their present value is great, even in contemporary American economic terms, and that it has risen greatly over the past few decades.

The federal agencies suddenly have become responsible for a vast pool of capital which they have been ill-prepared to manage. The management problems of this capital pool will surely intensify in the future. The federal land agencies and their clients should expect that other interest groups will increasingly cast longing glances at this capital, which up to now has generated meager returns.

Major Alternatives for the Future

If growing dissatisfaction with present public land management does lead to major changes, then it would be helpful at this time to consider the possible alternatives. As I contemplate the history of federal lands and speculate on their future, five broad alternatives for management of these lands seem possible:

- Most or all of the present federal lands will be retained in federal ownership, but strenuous efforts will be made to improve their management, and these will be more concern for the costs and returns from such ownership and management.
- Most or all of the present federal lands could be turned over to the states, without charge or on payment of some price, with their future management or disposal to be determined by the states.
- A major part or all of the present federal lands could be sold to private individuals, corporations, and groups or associations, under terms which would be spelled out in the enabling legislation.
- All or large parts of the present federal land area could be transferred to public or mixed public–private corporations, to be managed as decided upon by those corporations, but under terms of enabling legislation.
- Long-term leasing of federal lands could be extended greatly, not only as it has been done for mineral leasing but also for other commercial uses, and for conservation or preservation purposes as well.

Each of these five general alternatives will be discussed in more detail in the sections which follow, but first some general relationships should be noted. No one of these alternatives is likely to be the complete answer, a mixture of changes seems more probable. Most of the land could remain in federal ownership, as at present, with only modest disposals to states and to private parties and none at all to public or mixed public–private corporations. Or, at the other extreme, most could be sold to private parties with only a little remaining in federal ownership. Or any one of many other intermediate steps might be put into effect. In the sections which follow, each major alternative is discussed as if it were the sole line of change. But, in fact, each could be combined in almost any proportion with the others—allowing, of course, for the fact that the total federal land area sets an overall constraint on the number of acres that could be allocated to any combination of management systems.

In addition, each of these alternatives, if they involved less than the total federal land area, would be selective in character. The disposal and reservation processes of the past were highly selective; the private parties and federal agencies each chose the best land available to them, given their knowledge of land capabilities and the economic circumstances of the time. This selectivity will continue, again within the limits of land availability, knowledge, and competition from other management systems. Even if all, or essentially all, the present federal land remains so, the management practices applied to it will be selective— more on some types of land and less on others. If given freedom to make choices, all those choosing some federal land for purchase or for lease will choose only the land best suited to their needs.

Each of these major alternatives for future federal land management will require time to work out fully. Greater efficiency in federal land management will take some years to achieve. Any sale or lease disposal efforts will take some years to be carried out. Even if a national commitment to sell off the federal lands were acceptable to the majority of citizens—a most unlikely set of circumstances—it would take years for the disposal process to run its full course. Thus, regardlesss of which major alternative or combination of alternatives are chosen, and regardless of the vigor with which such lines of action are pushed, a considerable amount of land presently in federal ownership will remain so for a good many years.

Retention with greatly improved management

There are many ways in which the management of the federal lands might be "improved," although specific measures that constitute improvement to some users are not necessarily favored by others. The basic issue is to provide the maximum benefits on a long-term basis in relation to the costs incurred. Ideally, for every sensible management unit within the federal estate, the values or benefits (including those not customarily marketed for cash) should have the most favorable possible relationship to the costs (including, for example, interest or other capital charges on the value of the property), and the resultant management system should preserve the productive capacity of the area. If this ideal is achieved, then no management practice can be added, subtracted, or altered without lowering the net total benefits. Application of this standard does not require an allocation of costs to each resulting output nor does it require an exactly even flow of output each year, if changes in inventory values are properly evaluated.

Some aspects of this standard may require elaboration. The values or benefits should be the full value to society of the output. For those goods or services from the federal lands which are sold for cash—timber, grazing, oil, gas, coal, and other minerals—the present prices are often below fully competitive market prices, and the best estimates of the latter more accurately reflect value than the prices actually received. For those goods and services from the federal land not ordinarily sold for cash—water, wildlife, recreation, wilderness, aesthetic values—it is necessary to make some estimate of the value produced by the provision of these goods and services. Even those who profess scorn for economic analysis choose some outputs more than they do others; hence they have made an economic comparison, however crude it may be. In some cases, a judgment can be made without specific numbers that ranks one value above another.

In addition to including all cash costs, estimates must give some consideration to alternatives forgone by one method or objective of management as contrasted with another. If one system of management includes road building and another does not, then total costs obviously differ. Nonetheless, it is not necessary to make a cost allocation to each output in order to contrast the net results of one combination of outputs with another. The costs also must include reasonable charges for the use of the capital involved. Today, the federal lands have great capital value, and their economic management must return reasonable interest on this value. Indeed, one of the best expressions of values forgone by any management program is the interest on the capital value that would exist from some other system of management.

This standard of economic management should be applied to sensible management units, as is evident in the field. Within a national forest or grazing district, one management regime would be followed on sites with high productivity—whether that productivity was for timber growing, wilderness, or any other purpose—while a different regime of management would be taken for sites of lower productivity. But there obviously would be spatial limits to such management differentiation—for example, a 100-acre tract might be set aside for different management when a 1-acre tract could not reasonably be separated from the surrounding area.

Last, this standard of economic management should be applied with due consideration to cost of analysis, to ease of understanding by the users and by agency personnel, and to time required for the management analysis. Actually, a rough estimate may be as accurate as a finely tuned one if the uncertainties about the future are great. Management planning involves costs, and those costs should be included in the analysis.

There is good reason to believe that the Congress, in passing the 1974 Renewable Resources Planning Act, the 1976 National Forest Management Act, and the 1976 Federal Land Policy and Management Act, was striving generally, but not precisely, for just the kind of standards described above.[5] This is the general conclusion of Krutilla and Haigh:[6]

> The Renewable Resource Planning Act as amended by the National Forest Management Act can be looked upon as the culmination of a process of changing the congressional directives to the Forest Service from those consistent with an initial custodial role, to those of management—and indeed ultimately an intensive management—role in the administration of the national forests. . . . There appears to be little doubt that Congress has charged the Forest Service to manage the forest and rangeland resources in an economically efficient manner. . . . The objective of such a management approach is to maximize the aggregate value of the beneficial uses of all the resource services of the national forests without impairing the long-run productivity of the land.

All three acts stress the planning process and require statements of objectives and public involvement and interdisciplinary teams, in particular. The 1974 Act requires in Section 4 (1): "Specific identification of Program outputs, results anticipated, and benefits associated with investments in such a manner that the anticipated costs can be directly compared with the total related benefits and direct and indirect returns to the Federal Government."[7] The 1976 Act [Section 6 (g)(3) (A)] ensures "consideration of the economic and environmental aspects of various systems of renewable resource management, including the related systems of silviculture and protection of forest resources, to provide for outdoor recreation (including wilderness), range, timber, watershed, wildlife and fish."[8] Still other sections impose some constraints or restrictions on attainment of full economic management of the federal lands. It is true, however, that the 1974 and 1976 Acts, while requiring such balancing of costs and benefits in the *planning process*, only implicitly require that the resultant plans shall govern the actual administrative actions of the Forest Service. For me, the general thrust or objective of these three acts is fairly clear—more economic management of the federal lands—but I would characterize them on a whole as being less than precise, and somewhat ambiguous and or contradictory in some instances.

The final versions of these acts embody many compromises. The acts as passed are acceptable to interest groups only because of their vague language; more precise wording surely would have been rejected by some group. In the end, of course, the acts mean what the courts say they mean. It is sobering to realize that the full meaning of the 1897 Act providing (among other things) for timber sales from national forests was given final meaning by the courts only in 1974. It can well be argued that each act would have been more constructive if it had been passed many years earlier—for example, if the 1974 Renewable Resources Act had been passed in 1960 instead of the Multiple Use and Mangement Act, and if the Federal Land Policy and Management Act of 1976 had been passed instead of the Taylor Grazing Act. Congress normally responds only to a clearly felt need, and this, in turn, is one of the most serious limitations of federal land management.

If it is decided that the future direction of land management should be toward more economic management, we may well consider specific management or administrative steps to that general end.

One specific management practice toward that goal would be to establish a capital account covering the management of the federal lands, in total, or by specific agencies or administrative units (for example, each national forest). Under present procedures, annual appropriations for capital investment are lumped with those for current operations, a procedure that no private firm would ever follow or, in fact, be allowed by law to follow. Investments that should be productive over a considerable period of years are treated as operating costs for the year in which the expenditures were made. On the income side, no account is taken of the increase in capital values arising from these investments or from management practices. For instance, the national forests have been adding to their total volume of standing timber and at the same time have been slowly liquidating the largest old-growth timber. Is the value of all the standing timber rising or declining? No answer is readily apparent from available agency records, and the situation is enormously complicated by the great fluctuations in timber prices that have taken place in recent years. In addition, some of the investments are not based on appropriated funds. For example, often roads are built in the national forests by timber purchasers, and allowances for their cost are included in the timber sale contracts but do not show up as cash receipts in the revenue accounts.

Many of those advocating a capital account for the federal lands seem to assume that this practice would encourage larger appropriations for operations because the elimination of capital investment from annual appropriations would more favorably contrast appropriations with receipts. This, unfortunately, is totally erroneous. A true capital account requires annual depreciation charges on past investments and annual interest charges on the value of the resources. An estimated annual interest charge on the value of the capital embodied in timber and other separable resources is a reasonable charge against the continued management of the federal lands. If such capital costs were included, the federal lands as a whole—and the national forests in particular—would make a much worse economic or financial showing than they do when current appropriations, including capital investments, are compared with current cash receipts.[9] Nevertheless, an accurate economic analysis of the management of federal lands must include separate treatment of investments, of changing capital values, and of reasonable charges for the use of the capital.

Another practice that could facilitate more economic management of the federal lands would be to charge for some services, notably outdoor recreation, and to plow the revenues back into better management of these areas. Considerable public and political opposition has been voiced regarding the imposition of charges and fees for outdoor recreation on federal lands, even though they cover a substantial part of administrative costs of state and local parks.[10] One reason such charges are acceptable in state and local parks but opposed on federal lands is that the revenues have been used to provide necessary services for the areas that generated them. Earmarking revenues for specific expenditure is often opposed by political scientists and public administrators on the grounds that this frees the agencies from legislative oversight through the appropriation process. This objection easily can be met by the device of depositing the revenues in a special account from which the accumulated revenues can be appropriated each year by legislative action. This would give the users some assurance that their money is being used for their benefit, assure administrators of funds, and give legislators a periodic opportunity to control appropriations. If the cost of services provided bore some reasonable relationship to revenues generated, effective control on the level and suitability of federal management would be provided.

A more drastic step would be to limit operating expenditures on each unit of federal land (national forest, grazing district, and so forth) to total cash revenues collected from that area. This would make it difficult, if not impossible, to sell timber or other "goods" at a loss. It also would reduce sharply the present intensity of management on less productive federal areas but permit, if not encourage, intensive management on the more productive sites. If strictly applied, a requirement such as this would drastically alter the present situation in which a considerable part of the federal estate is managed at a loss. A milder version of this idea would limit operating costs to cash revenues plus estimated values of goods and services provided without cash return.

Some use of revolving funds or of multiyear budgets for the federal land-managing agencies would assure funds for longer-term management goals, thus reducing some of the uncertainties arising out of annual budgets. Aside from the political difficulties of getting legislative approval for multiyear budgets or revolving funds, there would be problems in anticipating needs for more than a year ahead, of having enough flexibility to meet natural disasters or other unusual situations, and of having funds available to meet changing needs. Longer-term appropriations could reduce inefficiencies arising from annual appropriations (which often are not made in timely fashion) but could do little or nothing about other inefficiencies.

Progress toward more efficient management of the federal lands since passage of the 1974 and 1976 Acts has been disappointing.[11] The planning process is under way, but thus far it has had dubious results. The appropriations to the Forest Service and BLM may have been increased as a result, but this measure of success is more agreeable to the bureaucracy than to the general public. The large deficits from national forest management have not only continued, but apparently grown, even in real terms. One defense of the 1974 and 1976 Acts is that it is still too soon to judge their results accurately. On the other hand, it may be argued that the goals of greater efficiency in federal land management are simply unrealistic. Perhaps the most accurate judgment today is that the jury is still out.

Unless efficiency is improved, the public's dissatisfaction with the present management will continue to increase and major changes will be proposed and, perhaps, carried out. If federal management should become as efficient as private management, has the rationale for continued federal ownership been weakened? Why not sell the federal lands to private parties? Conversely, would greater efficiency of public management weaken or remove the rationale for transfer to private ownership?

Finally, there is serious doubt that federal agencies, who are subject to the inevitable political pressures that affect public programs and who are caught up in the extensive bureaucratic and administrative web that encompasses federal activities, can ever hope to manage extensive lands and associated resources intensively and economically. Perhaps federal agency management of land was suitable for custodial management but

is less suitable for intensive and consultative–confrontation management
of the same lands.

Transferring the federal lands to the states

A second alternative would be to transfer the title to some or all of
the federal lands to the states in which they lie, either freely or upon
payment of specified, but presumably small, sums per acre, after which
time the states would manage the land. Nevada and Idaho had proposed
large-scale federal land grants to the PLLRC, prior to 1970, but their
proposal was rejected.[12] Some Sagebrush rebels advocate that the current federal lands be turned over to the states.[13] At a conference on
"Private Rights and Public Lands," sponsored by the Heritage Foundation and others on June 21, 1982, several speakers voiced this view.[14]

Relationships between the federal and state governments on land
matters have a long and complex history.[15] Never has there been any
significant acreage of federal lands in the original thirteen colonies, or
in the states formed out of them (Maine, Vermont, West Virginia,
Kentucky, and Tennessee), or in the states that had been independent
countries before admission to the Union (Texas and Hawaii). In all the
other states formed out of federal land (or public domain, to use a
common term applied to such states), beginning with Ohio in 1802 and
continuing across the Mississippi and Missouri basins, the South, and
the Rocky Mountain, Intermountain, and Pacific regions, to Alaska,
the matter of federal land grants to the states has been an important
political issue. In general, the following arrangements were made for
each state:

- Congress insisted and the states accepted a disclaimer relinquishing
all further claims to federal land. Some of the present-day advocates
of federal land transfer to the states seem to believe that this disclaimer
was forced only on the western states, the present-day public land
states, but, in fact, from Ohio onward, every state formed from public
domain has been admitted on the basis of such a disclaimer.

- The new states agreed not to interfere with the disposition of the
federal lands by the federal government and further agreed not to
collect taxes on the federal land. Sometimes, they also agreed not to
collect taxes for a few years following the passage of the lands from
federal to private ownership.

- On admission, the new states were given specific grants of land for
common schools. Each state admitted to the Union prior to 1850

received one section (640 acres) per township of thirty-six sections (23,040 acres). After 1850, this was increased to two sections per township, and beginning with Utah in 1896 it was increased to four sections per township. Nevada, which had received two sections when admitted in 1864, relinquished its grant of about 4 million acres and in return received a quantity grant for common schools of 2 million acres to be selected where it chose. Upon admission, in addition to the common school grants, the new states were given quantity grants, to be selected where they chose, for land grant colleges and other public purposes.

• The new states were allotted 5 percent of the receipts from disposition of the federal lands within their borders, to be used for roads or other purposes.

• Following their admission, the states formed from the public domain were given other grants, notably of swamp and overflowed lands. There was much fraud connected with the disposition of these lands, and a great deal of controversy arose between the states and federal agencies over these grants. The swampland provision was especially important in Florida, where title to more than half of the total land area went to the state under this law. It had a significant impact in Mississippi, as well as in Louisiana and Arkansas, who each received more than 7 million acres by this route, in Michigan, Minnesota, and Wisconsin, where many millions of acres were involved, and also in California where more than 2 million acres, including the most fertile parts of the Central Valley, were transferred under this law.

Although the forgoing provisions applied to all states formed from public domain, the exact terms of these general provisions varied. For example, in some states the common school grants were made to the states, to be administered as the state chose; in others, the grants were made to the township governments. Or, to use another example, the uses to which the 5 percent of federal revenues from public land disposal could be put varied somewhat from state to state. Nonetheless, for our purposes here, the general terms are sufficiently accurate and applicable.

Since the first grant of federal lands to Ohio in 1802, a total of 328 million acres has been granted to the states (table 10-1). This accounts for more than 14 percent of the total land area of the United States, for more than 18 percent of public domain, and for more than 28 percent of the total area of public domain disposed of. Clearly, grants to states have been an important channel for transfer of public domain out of federal ownership. The eighteen states not formed from the public do-

TABLE 10-1. FEDERAL LAND GRANTS TO GROUPS OF STATES FROM 1803
THROUGH FY1980
(in millions of acres)

| | Purpose of grant | | | | | |
Group of states	Common schools	Other schools	Railroads, wagon roads, canals and rivers[a]	Swamp reclamation	Other	Total
Eighteen states not formed from public domain[b]	0	5.2	0	0	—	5.2
Nineteen states formed from public domain, outside of West[c]	26.2	6.8	43.9	62.4	7.8	147.2
Eleven western states[d]	51.3	4.6	2.7	2.5	10.4	71.5
Alaska	0.1	0.1	0	0	104.4	104.6
Total[e]	77.6	16.7	46.6	64.9	122.6	328.4

Source: U.S. Department of the Interior Bureau of Land Management, *Public Land Statistics, 1980*
(Washington, D.C., Government Printing Office, 1980) table 4, pp. 6–7.

[a]This does not include the 94 million acres granted directly to railroad corporations.

[b]Connecticut, Delaware, Kentucky, Maine, Maryland, Massachusetts, New Hampshire, New Jersey,
New York, North Carolina, Pennsylvania, Rhode Island, South Carolina, Tennessee, Texas, Vermont,
Virginia, and West Virginia.

[c]Alabama, Arkansas, Florida, Georgia, Illinois, Indiana, Iowa, Kansas, Louisiana, Michigan, Minnesota, Mississippi, Missouri, Nebraska, North Dakota, Ohio, Oklahoma, South Dakota, and Wisconsin.

[d]Arizona, California, Colorado, Idaho, Montana, Nevada, New Mexico, Oregon, Utah, Washington,
and Wyoming.

[e]Items may not total exactly because of rounding.

main received only small grants of federal land, and that was in the form
of scrip because they had no federal land within their borders. These
grants were primarily for the establishment of land grant colleges. The
nineteen public domain states outside of the West received large grants
for a variety of purposes: a substantial part of their grants was swamp
and overflowed lands, but grants for railroads, wagon roads, and canals
were also extensive. Their common school grants made up less than
one-fifth of their total. The eleven western states also received large
grants, and two-thirds of these were common school grants. Of the
western states, only California received significant swampland grants.
By the time these states were admitted to the Union, most grants for
railroads were made directly to the railroad corporations, so the grants
for this purpose were relatively small in the West. If direct grants to
railroads in the West had been made as grants to states which in turn
could be granted to railroads, as was common in the Midwest and South,
the total grants to western states would have been much larger. Finally,

Alaska was the recipient of a huge area of federal land, far more acreage than any other state, but a smaller proportion of its total area than Florida received.

The data in table 10-1 are for lands whose titles have passed from the federal to a state government. Dragoo notes, "In addition to those grant lands that have passed title to the states, large acreages of public domain are still outstanding to several states. Many of the state entitlements still owing are in lieu selections to replace lands which were never received by the states due to federal withdrawal or third party withdrawal pursuant to public land laws in effect at the time of statehood grant."[16] In 1970 the PLLRC noted that 900,000 acres were still due the states.[17] Dragoo says that, in 1975, 200,000 acres were still due to Utah and 180,000 acres to Arizona, the two states with the largest area of unsatisfied grants. In any alternative program established for the future administration of the federal lands, it is necessary to provide for satisfaction of these grants.

The large federal land grants to the states did not satisfy their appetites and, indeed, may have whetted those appetites for additional lands. As Gates reports, "The states were never to accept Federal management of the lands and Federal reduction of their power with equanimity and were to devote much time to considering plans for reversing the disclaimer provisions in their enabling acts and ordinances they were compelled to accept."[18]

As their population rapidly increased, the public domain states of the Midwest and South eagerly converted federal land to private ownership. And although they were eager to change the disclaimer provisions in their enabling legislation, they lacked the political power to do so. Today, they have nothing to gain from changing their enabling acts, and there is little interest in pursuing such a goal. Indeed, if an effort were made to revise disclaimer clauses, they would be more likely to line up with the older states in opposition to more favorable treatment for the western States. As John Francis of the Brookings Institution has pointed out, there is not even unanimity *within* the western states on the issue of federal land transfer to the states.[19]

Gates clearly shows that the disclaimer provision in the state enabling acts was no barrier to further substantial federal land grants. Most of the swamp and overflowed land grants and the transportation grants to the states (see table 10-1) were made after the disclaimer clauses had been imposed on the affected states. If there were a substantial national consensus for major new federal land grants to the states, the disclaimer clauses in their enabling legislation would not prevent them from being awarded. This is not to say, of course, that states outside of the West would generally support such further federal land grants.

How well have the states managed the federal lands granted to them? Drawing on the available record, Gates comments, "The early story of state management of public lands is not a pretty one."[20] Incompetence was mixed with fraud in most states, and money derived from the sale of grant lands was frittered away in many instances. Gates feels that state administration of grant lands has improved greatly in recent years, noting that "gradually an increasing sense of responsibility in public administration and a wider interest in the purposes for which the grants were given contributed to improved management of state lands."[21] Some of the sources he cites do not quite support his optimism. He does point to the very real accomplishments of New Mexico and Washington in retaining much or all of their grant lands and in making them yield substantial revenues for the school systems. And he winds up on an optimistic note, "In fact, one can say that the newer states have done as well as has the federal government in the management, sale, and leasing of public lands."[22] My own verdict—based on a considerable knowledge of state land administration over the past four decades—would not be so generous, although I too agree that major progress has been made.

There would be almost certain opposition from conservation and other groups in the West and elsewhere if a major effort were mounted by Congress or the administration to turn major additional areas of land to the states, regardless of the terms proposed for the transfer. Some opposition would be based upon a fear of state incompetence or mal-feasance, and other opponents would express fears that the states would have a greater interest than the federal agencies in cash returns and less interest in preserving resource productivity for the future. The belief that a state administration would be closer to users would argue both for and against a larger state role. Questions would also be raised about the competence of the states, as compared with that of the federal agencies, with some critics contending that substituting state for federal administration would be the worst of both worlds; that is, it would remain public, not private, and under state rather than federal control. Of course, opposition inevitably will be voiced toward any future man-agement scheme for the federal lands, so that the probability of op-position should not be taken as definitive in arriving at decisions about future management of the lands now in federal ownership.

Disposal of federal land to private ownership

A third alternative for future management of the federal lands cur-rently administered by the BLM and Forest Service would be their disposal, by sale or other means, to individuals, corporations, and groups of all kinds.

The virtues of private, as contrasted with public, ownership and management of land have been described by Gardner in chapter 8, and by Libecap, Baden, Stroup, Hanke, and others.[23] The transfer of land from federal to private ownership has had a long history. The total area of land ever in the public domain amounts to more than 1.8 billion acres. By 1980, 1.2 billion acres, or 63 percent of the total, had been disposed of by one means or another to some class of private recipients, and this was in addition to the grants to states discussed earlier.[24] Of these disposals, 287 million acres had gone to homesteaders, 94 million acres directly to railroad corporations, 61 million acres as bounties to war veterans, 70 million acres by a variety of miscellaneous acts (timber culture, desert land, and others), and 304 million acres by methods described as "not elsewhere classified," which in practice means nearly all by competitive sale. Thus, more than 800 million acres, or 35 percent of the total land area of the nation, or 45 percent of the area ever in the public domain, or 71 percent of all land transferred out of federal ownership went directly to private ownership. The states sold or gave away to private owners much of the land that had been granted to them.

These figures substantially understate the importance of past transfers of land from federal to private ownership. These transfers took place despite the fact that no legal means existed for the transfer of forested and grazing lands in economically sized units, particularly in the West where the minimum size of economically viable ranch and forest enterprises was large. Moreover, they took place despite the substantial withdrawal of land, beginning in 1891, for the forest reserves (now national forests) and for other purposes. Disposals had continued at a rather rapid pace all through the 1920s and early 1930s, contrary to a widely held belief that large-scale transfers of land had *de facto* come to an end by 1934 when the Taylor Grazing Act was passed and all remaining federal land was withdrawn from disposal to private ownership except upon classification and approval by the Department of the Interior. It is true that virtually all land capable of cultivated crop agriculture had been homesteaded or otherwise disposed of by 1920. But total entries for federal land ranged from 3 to 5 million acres annually from 1923 (when the flush of homesteading for crop farming had ended) until 1934.[25] These disposals occurred despite laws particularly unsuited to the kinds of land involved and to the kinds of uses contemplated. But this rate of disposal, continuing unchanged for forty or fifty years, would have exhausted the unreserved public domain as it stood in 1934. One can only speculate, of course, as to the acreage that might have been disposed of by now, had the Taylor Grazing Act not been passed in

1934. But my judgment is that the area would have been considerable. Likewise, had the national forests not been established when they were, substantial parts of them by now would have been in private ownership.

Those who state that the federal lands, the lands managed by the BLM, and certain national forests are merely the scraps of the public domain and unsuitable for private ownership (the "lands nobody wanted") are implicitly accepting a totally static view of history and economic change. Land on the frontier always has had a marginal value—in fact, that was one definition of the frontier. As settlement progressed and the economy developed, land which previously had had little or no value came to be more valuable. Elsewhere I have described the disposal of the federal lands as a modern miracle of loaves and fishes—the more federal land given away, the more valuable became that which remained.[26] The same economic changes that have increased the total value of the federal lands more than twenty-fold since the 1920s have equally increased the values of the intermingled private lands and of the remaining federal lands. Checkerboard lands granted to railroads in northeastern Nevada, which could have been bought from the railroad land company for $2.50 per acre as late as the end of World War II, are now selling for $50 an acre, and ranchers who earlier hesitated to make investments in grazing land are now convinced that they must own more of the lands on which their livestock graze. Similarly, in the interwar years timber processors, who owned little forestland and bought their timber from public or private landowners, preferred to invest their limited capital in plants and equipment. Since World War II they have actively sought to acquire the land from which their timber must come. Many small firms (and some large ones, too) have been consolidated during the last thirty years into larger, vertically integrated forest firms. Had the means existed whereby such firms could have acquired federally owned forestland, it seems clear that many would have done so.

The conservation and preservation organizations have generally sought to use their economic and political strength to secure the establishment of national parks, monuments, and wilderness areas. In addition, some of them also have acquired land by purchase or by gift from other private owners. The Sierra Club and the Wilderness Society have not sought the role of landowners, but the Audubon Society and the Nature Conservancy have acquired considerable areas by purchase, gift, or bequest.

Had laws existed under which organizations could have secured federal land in units and at prices suitable for the uses each sought to make of them, they too might have acquired substantial areas of federal land in the past. Private ownership does, after all, provide the user greater security of land tenure than does any system of management for the public lands.

My judgment is that substantial acreages of today's federal lands would be purchased by ranchers, forest industry firms, mining and oil firms, and conservation groups if they were given the opportunity to do so. Much would depend, of course, on the sale terms. Price per acre is not the sole factor, however. The eagerness with which private parties would seek to buy would be influenced by the minimum and maximum units of purchase, the constraints on the future use of the land, what preference would be given to present users or to owners of adjoining private lands, and general economic conditions. The rate of sale might vary, from relatively rapid to relatively slow. Competitive sale is one obvious means of disposal, but so-called competitive sales may not be very competitive if there is only one potential buyer, as the sales of timber from national forests and O&C lands have demonstrated.

The Reagan administration has announced its intention to sell "unneeded" federal lands and properties and to use the receipts from such sales to reduce the total federal debt. All statements on this subject have emphasized that there is no intent to sell national parks, wildlife refuges, wilderness areas, scenic river areas, and similar conservation areas. The proposal has been covered in countless press releases and media accounts,[27] but all the publicity relates to preliminary plans, when, in fact, the reality well may be different. Great emphasis has been placed on the fact that the lands proposed for sale are those which the agencies do not wish to retain—or, at least, there are others that they have a greater interest in retaining.

The BLM has identified some 4.4 million acres which can be legally sold; the Forest Service has put 6 million acres under study, some of which it may decide should not be sold, and for which it would have to obtain legal authority to do so. These BLM and Forest Service lands are in addition to other areas, some within cities, which other federal agencies are willing to sell. Sales of these lands would be similar to the garage sales of suburbanites—offering unneeded articles, perhaps useless ones, in the hope of raising a little cash.

It should be pointed out that the BLM has "sold" or is in the process of "selling" much of its land, including Outer Continental Shelf areas, as part of its long-term oil- and gas-leasing program. Although running for a specific period, they remain in effect as long as oil and gas is produced in paying quantities. A lease finally expires when there is little value remaining. This is true especially for the Outer Continental Shelf areas; and while these leases include only one use of the land, often that use constitutes by far the greater part of each tract's total value.

Something similar exists for coal leases. Grazing permits generally cover ten years but usually are renewed for longer. Politically, but not legally, ranchers have a substantial claim to tracts of both grazing district and national forest land. Wilderness buffs also have strong political claims to established wilderness areas.

Much is made of the possibility of reducing the federal debt by use of the sales receipts from these lands. This may be good public relations, but it has limited economic significance for the scale of the proposed land sales. If the BLM is successful in making its projected sales and if such sales do, in fact, yield the estimated revenue, this amounts to little more than 2 percent of the total federal debt. Moreover, there is no assurance that the total debt will actually be lowered by such receipts from federal land sales—the debt responds as well to other economic and political factors and the relatively small revenues from land sales may be totally overwhelmed by other events.

The federal lands as a whole have great value and a substantial private market exists for them; but to realize any large return, the federal government would have to sell portions of the most valuable forests, grazing lands, and mineral lands. If debt reduction is a serious objective of federal land disposal, then such disposal must go far beyond any lands currently being considered for sale. Disposal on such a large scale would have to be at the option of buyers who would nominate tracts of interest to them; the agencies would never on their own propose to sell truly valuable federal lands.

The national forests as a whole and most of the grazing land administered by the BLM costs more for current administration than the revenues received from sale of timber, forage, and other products except oil and gas, and from provision of services such as recreation, wildlife, watershed, and wilderness. The federal Treasury thus would have an increased current cash flow if these unsold lands were given away. I would emphasize that, at the least, the relationship of current costs to current revenues should be a factor for consideration.

Public or mixed public–private corporations

A fourth alternative for the future management of all or most of the lands now federally owned would be to turn their management over to a federal corporation(s) or to mixed public–private corporations.

Public and mixed public–private corporations have been established for various purposes in the United States, but to date none have been established for federal land management. Where the functions to be performed are of a commercial or semicommercial nature, with substantial investments, expenditures, and revenues, and with substantial dealings with a user public which involve payments for use, a corporate form of organization may be quite appropriate. Intergovernmental organizations, such as the New York Port Authority, have the attributes of a corporation, and are one way of pooling governmental powers from separate units of government. There is no need here to describe and analyze these various corporate or quasi-corporate arrangements, except to note that they do exist, but not for the federal lands.

Because up to now the land-managing agencies have had no direct experience with public or public–private corporations, it is impossible to draw guidance from past experience. Instead, we must rely on analysis of proposals conceived by their proponents. While such a course is inevitable if anything new is to be developed, the possibility exists that an important problem has been underestimated, that some feature will not work as estimated, or that once in operation some unforeseen hitches will develop. And often an innovation may not prove sound once it has been put into practice. Nevertheless, some proposals have been made, and this is one alternative for the future that should be explored.

In 1957 Held and I outlined the possible structure and functioning of a federal land management corporation, at one of the first times such a proposal was made.[28] Our concern was to secure greater efficiency in federal land management by offering more incentives to the land managers and greater freedom to invest in federal lands when economic circumstances so warranted. As we said:

> From an economic viewpoint, the basic need for federal land management in the future is for a more careful appraisal of expenditures and revenues; for larger appropriations, especially for investment, in order to make the lands fully productive; and for generally less restrictive methods of resource management. At the same time that these basic needs are met, the methods of land administration must preserve the nonmonetary values of the federal lands, and must also be compatible with the American system of government. The administration of these lands must be not only responsible but also responsive to the wishes of the public.[29]

Our proposal was for a single federal corporation to encompass all lands managed by the Forest Service, the BLM, the National Park Service, and the Fish and Wildlife Service. We believed that a corporation managed by professional career employees would deal more generously with outdoor recreation, wildlife, and other nonmonetary values than Congress and the executive budget process had dealt with them in recent decades. Comparatively, there was less concern then over wilderness areas than there is today, and for this reason we did not explore wilderness designation and management as fully as other aspects of federal land management. At that time, the national parks had been seriously neglected and, as a result, many facilities were sadly run down. Very little of an explicit nature resulted from our suggestion. In 1958 the Senate's Interior and Insular Affairs Committee (as it was then known) made a tentative move toward requiring the federal agencies to provide a consolidated annual investment, income, and expense statement, such as a private corporation normally provides, but this was never followed up and came to naught. Our proposal may have had some influence on later legislation, but this would be difficult to prove. In the light of twenty-five years' experience since our proposal was made, it now seems to have been rather naive. The interagency rivalries in such an all-encompassing federal corporation would have been very great. We gave little consideration to the possibility of substantially different management programs for highly productive and for poorly productive lands, though that might have been resolved later. However, we were struggling to describe existent problems and to suggest major solutions, and the legislation of 1974 and 1976 clearly was aimed at the same problems.

In 1981 Nelson outlined a program for management of all the federal lands in the future.[30] His program is notable in how it differentiates types of land and their expected uses. He envisages only modest changes for the National Park System and for the Wildlife Preserve System, but he would transfer large areas of recreation lands of primarily state or local interest to the states. He does not specifically say, whether this would be by sale or by gift. For the recreation lands of national interest outside of the national parks—the wilderness, conservation, and wild and scenic areas, particularly—he would have a Federal Recreation Service—perhaps created from the current Forest Service—assume management. He would lease grazing lands "to ranchers for grazing purposes for 50 to 100 years at a time, with the preferential option to renew."[31] He would turn over leasable minerals to the states but leave the mining law more or less intact with its administration in the hands of the agency managing the surface resources of each area. This is a very brief and somewhat sketchy review of a carefully reasoned paper.

For the purpose of this section, attention should be focused on Nelson's proposal for the 40 million acres of national forests and the 2 million acres of O&C lands which he designates as "prime." This is a somewhat larger area than I classified as "commercial Class A" forest.[32] Whatever may be the exact acreage, we agree that only a part of what the Forest Service statistics show as commercial forest has truly economic possibilities for sustained forest management. As Nelson says:

> It is possible that the Forest Service could be transformed to show a much greater capacity for a businesslike, efficient approach to timber management. . . . A much greater likelihood of introducing efficient timber management to a limited acreage of high quality timber stands is found in the creation of new timber management corporations. Initially, stock in such corporations could be distributed to match the current distribution of revenues from public timber operations. Local counties thus would receive 25 percent of the stock in timber corporations created from national forest lands, with the remaining 75 percent of the stock going to the Federal government. Federal shares would then be sold gradually over a transition period. At the end of the transition the ultimate objective would be a private timber corporation—although the local counties could of course retain their shares.[33]

Although Nelson speaks of "corporations," he provides no specific number of such corporations or the acreage each would average. Neither does he provide specific information as to the probable length of the transition period or as to how the federal shares would be sold to private parties. Nor does he discuss in detail why he proposes the mixed public–private corporation when the ultimate goal is a private one. Despite these considerations, Nelson's proposal is clearly an imaginative and innovative one. The missing details can be developed from debate and further public consideration.

In 1982 Teeguarden proposed the creation of public corporations, apparently for all federal lands, although all his specific illustrations are for the national forests.

> . . . I outline yet another alternative: the establishment of independent public corporations which could function like investor-owned public utilities. For example, each national forest could be chartered to operate as a public corporation, with a Board of Directors and administrative officers. Each unit would have the legal right to establish its own timber production and other resource output goals in response to the demand for its services and costs of producing them, just as do private investor-owned utilities. Operations would be self-financed through sales of products and services, and land-use leases. Investment capital would be obtained by the sale of tax exempt bonds or, as in the case of private corporations, by retained earnings. Services that would not voluntarily be provided because of lack of adequate profit margins or markets, but which were considered to yield significant public benefits, could be assured through contractual arrangements in which agencies of the Federal or State governments pay the National Forest Corporation for specified kinds of quantities of such services.[34]

Teeguarden's other provisions include a Public Corporations Board, for general oversight; charters of limited duration, say, ten to fifteen years, which would not be renewed if the corporation could not earn sufficient profit to maintain its operations; a requirement that the corporations earn 11 percent return on their assets; requirements for reporting financial operations annually; and suggestions for splitting any profits. He continues: "A system of public corporations, operating under the supervision of a Public Corporations Board, is one of several alternatives to the present method for managing the federal lands. Obviously many more details need to be developed before the proposal can be fully evaluated. My intent here is merely to stimulate serious study of what seems to be a potentially promising way of achieving more effective management of federal lands."[35]

He goes on to suggest that "consideration . . . be given to designating a portion of the federal lands, say 25 percent, including national forests, for management by public corporations."[36]

Teeguarden, as many another observer of current federal land management has been, is concerned with certain basic problems: the need to simplify a system of land management which has grown burdensomely detailed and complex; a need to impose some system of economic accountability on the federal land managers; and the need to devise a system more locally adaptable than a national management system. His proposal, tentative as it is, is unclear on several points: Are national forests as now delineated too small for the kind of efficiency he strives for? How will mineral leasing and mineral disposal be handled? Is the rate of return that he postulated for capital unrealistically high? If there are extensive areas of federal land which cannot be self-sustaining, let alone return 11 percent on their capital value (as I strongly suspect may be the case), how would these be handled? But these specific questions are not meant to detract from the value of his proposal.

For any of the proposals described in this section, or for any others that might grow out of them, new legislation would be required. Its specific terms and limitations clearly would be critical. Even if agreement could be reached for a trial program or for a program of general applicability, and even if corporations were set up which initially were quite successful, would this really be a long-term solution to the problems of efficiency in federal land management? I am reminded of the enthusiasm with which the Tennessee Valley Authority and the Bonneville Power Authority were created in the 1930s, followed by their rather successful operations for some years, and their increasing bureaucratization and inertia to meet today's changing times. A new broom sweeps clean, but how long does it last?

Large-scale, long-term leasing of federal lands

The fifth and final alternative for future management of the federal lands is large-scale, long-term leasing of these lands for conservation and preservation uses, as well as for commercial uses.

Leasing, as contrasted to ownership, of land and associated resources is a time-circumscribed transfer to the lessee of some parts of the rights which go with land ownership, for a specified period of time, for agreed-upon uses and methods of management. Leasing is very common for privately owned land and its associated resources. More than one-third of all residential units (houses and apartments) are occupied by renters, for instance. Much farmland is also rented, sometimes as whole farms and sometimes as additional acreage to supplement the land of an established farmer. Likewise, much office and commercial space is rented. But renting extends also to minerals development—a considerable part of all oil and gas development is under lease, by one private party from another. Some forestland is also leased from private landowners, often small ones, by forest industry firms.

This historic experience with private leasing of many kinds of property, in the United States and other countries, has established certain principles governing leases that are equitable for both the lessor and lessee, and lead to efficient use of the property. First of all, the lease should be written, not oral; while it may seem inconceivable that a federal official and a private party could or would enter into an oral lease for federal land, it is not inconceivable that one or the other might make some verbal statements that lead the other to expect some performance which will not be forthcoming. If disputes arise, the written lease will govern. The lease also should preserve as much flexibility as possible for the lessee to make economically efficient use of the property. The lease should spell out what can, and what cannot, be done; and it should provide the lessor a reasonable opportunity for inspection but under circumstances that are not unduly inconvenient to the lessee.

Well-drawn leases provide for consideration of their renewal. In general, renewal negotiations should begin when about one-fourth of the original lease remains—three months before the end for an annual lease, two or three years before the end for a decade-long lease, twenty or twenty-five years before the end of a ninety-nine-year lease, and so forth. If they cannot agree on an extension, this gives both lessee and lessor time in which to work out other arrangements. The lessee should either agree to pay for specified improvements on the property when the lease period begins or should agree to maintain improvements at a specified level, or both. The lease well might include penalties for degradation of the property. A performance bond, to ensure that the lessee lives up to the terms of the lease, could also be included, perhaps in the form of Treasury notes (suitably rolled over, as they matured) from which the lessee could collect the interest, thus substantially reducing the annual cost while at the same time leaving the government adequately protected. The performance bond obviously should be large enough to guarantee performance in the event of failure of the lessee to carry out some agreed-upon action.

The lessee should also provide the circumstances, if any, on which either the lessor or the lessee could cancel the lease before the end of the term, and the penalties that should be exacted for cancellation. The lessor should agree to compensate the lessee for the unexhausted or undepreciated value of the improvements at the end of the lease period. This provision, plus the provision for negotiation over renewal, should rule out neglect on the part of the lessee to maintain the property, which otherwise might occur near the end of a lease period.

Last, the well-drawn private lease provides for the time and form of payment of rent—cash, share, or cash–share. It should also provide for the sharing, if any, in cost items.

The term *lease*, as it is applied to federal land has a specific legal meaning; in addition to arrangements called leases, there are other authorized temporary uses of federal land called permits, licenses, or some other term. For the discussion which follows, I use lease to include all such arrangements whereby individuals, corporations, or groups are allowed to make some time-limited use of federal land for some defined purpose under specified conditions or terms, with the lessee having substantial discretionary power over the actual use and management of the land.

As thus defined, leasing of federal land is much more common than often is realized. The family which picnics at a national forest or BLM site has exclusive use of a small area, for a specific purpose, for a limited time; they are not permitted to cut down trees, drill for oil, graze cattle or sheep, or do any of many other things permitted on some federal land under some conditions. When this family has finished its picnic, it yields its site to some other group for the same purpose. The person or group which camps for one or more nights on federal land is in a similar situation except that they spend a little more time and engage in a wider range of activities.

The rancher who has a ten-year permit to graze livestock has a longer tenure, subject to rules and regulations of the federal agency, but again for a specified purpose. Although his permit is for ten years, usually they are renewed so that ranchers have in practice, but not in law, longer tenures. Because their permits in theory may not be extended and be-cause in practice the number of animals or season of use may be changed by the federal agency, their tenure is limited and insecure. The oil and gas lease is in effect for a specified number of years, but thereafter it runs as long as oil and gas is produced in paying quantities—and this may be for many years. The timber purchaser has a defined time in which to remove the timber and to perform any site restoration which had been specified in the sale.

When this inclusive definition of leasing is used, it is apparent that nearly all the federal land is under some form of lease and that the number of lessees is very large, about one-fourth or more of the total population of the country. What is envisaged here is a major expansion of leasing in the more formal sense of the term, for much longer time periods, and under terms for the various uses which are both nearly uniform but at the same time flexible enough to meet the needs of lessees for different uses.

Leasing for all these and other uses could be greatly extended to private parties and the terms and conditions of all leases could be more carefully designed to meet the objectives of the user groups interested in such lands. There is no logical reason why recreation areas and wil-derness areas could not be leased to some groups. There have been a few instances of leasing municipal watersheds on federal land to the cities whose water supply originated there.

In addition to following the general strictures commonly used for private leasing arrangements, there are other points which apply spe-cifically to the leasing of federal land. The lease should not exacerbate the problems of intermingled federal and nonfederal land ownership. On the contrary, it would be helpful if those problems could in some degree be reduced by the leasing. If the owner(s) of intermingled check-erboard land were given preferential rights in leasing of federal land, the whole area could be blocked up into more efficient management units. The lessee would be held to the same environmental controls that apply to privately owned land, although the issuance of the lease would not require that the lessee prepare an Environmental Impact Statement. It also should be noted that the leasing of federal land to a private group does not affect the state's legal authority over any game animals on the land.

The basic goal in formulating terms for private leasing of federal land would be to harness the self-interests of the lessee to the social goals of the national government as landowner. To the extent that the terms of the lease could harness those self-interests, the ingenuity and ability of the private lessees would be enlisted and supervision by the federal agencies could be reduced. Violations of the terms of the leases would also be less likely to occur, and the lessees would surely find many ways of managing the land resources more effectively.

Clearly, the terms on which leases of federal land would be available must be made attractive to the prospective lessees, otherwise they will be unwilling to enter into a lease. Similarly, the terms must be acceptable to the federal agencies and the electorate as well, if they or the agencies are to have discretion in granting or refusing lease applications. Present laws would have to be amended if leasing were to increase. The degree to which the terms of leases would be spelled out in the legislation and the degree to which the agencies could be granted discretionary decision-making power must also be considered. For instance, at present both the Forest Service and BLM are required (with limited exceptions) to offer timber at "competitive sale," but the specific decisions about the sale are left to the agencies. Or, in another example, the royalty rate for oil and gas leases is specified in law, but the decision to offer a specific tract on the Outer Continental Shelf resides with the secretary of the interior. At one extreme, prospective lessees could be given the right to demand a lease under terms specified in law, without the agencies having power either to refuse the lease or to modify its terms; at the other extreme, the federal agencies could be given nearly total discretionary power over leasing; and, of course, innumerable intermediate alternatives exist. If the idea of more extensive, regularized leasing of federal lands is taken seriously, one can be sure the specific terms of the law will be debated at length.

As outlined here, long-term leasing of federal lands would give lessees many of the advantages of full ownership of the land and resources: for a defined time, subject to defined and agreed-upon conditions, the lessee would have the opportunity to manage the land and resources as the lessee chose—for profit if for a commercial use, or for other objectives if the lessee wished. Federal expenditures for land management could be drastically reduced, nearly to nothing, for the leased land. Toward the end of the lease, the federal government would have the option of managing the land itself, of renewing the lease, or making arrangements for other disposal.

An Illustrative Proposal

The forgoing general ideas about increased regularized long-term leasing of federal lands may be more readily understood if presented in the form of a specific proposal. General ideas sometimes sound attractive and practical, but flaws are revealed and criticisms arise when they are put into operation. The proposal in this section is only one of several that might be developed. It includes provisions for each major use and user group. Although it is only an illustrative proposal, in my judgment it constitutes an economically and politically realistic alternative.

First of all, all leases issued under this proposal would be subject to the same terms as those found in any well-drawn public or private lease. Second, the illustrative proposal would open all federal lands to lease, but I would judge that the total area likely to be leased under this scheme would not be much more than one-fourth of the federal lands in the "lower 48" states. The remaining federal lands would be managed under one of the other alternatives I have outlined. If there were a national consensus to lease more of the federal lands, it would be necessary to establish different terms.

Third, leasing would be voluntary. A prospective lessee would seek a lease only when it would improve his or her position. Groups, companies, individuals, and units of local government could feel pressured to lease a tract of federal land, but would do so only if they felt their position was stronger with a lease than without one. If the unselected federal lands were continued under present federal management, then all present laws, regulations, and procedures applying to those lands would continue. If they were offered for sale or transferred to states, then applicable laws would govern their management.

These terms have been carefully chosen to make only the most productive federal lands attractive for leasing—whether productivity is measured in terms of ability to produce wood or grass or oil or to provide recreation, wilderness, or wildlife values. In its own way, the leasing process will be as selective as the disposal and reservation processes of earlier decades.

Lessees operating under the terms of this specific proposal would not be limited to a single lease. At present, oil and gas and other minerals-extracting companies often hold more than one lease, and timber buyers often have more than one purchase unit under contract at any date. The possibility of multiple leases would be extended to any and all lessees for all kinds of land and for all forms of use. The minimum acreages per lease that have been chosen are large enough to be significant for any prospective lessee but not so large as to exclude smaller, economically viable operations. The minimum acreages are not by themselves economically optimum, but have been established to discourage nuisance filings and to avoid fragmentation of the federal land. There would be no maximum acreage for a single lease under any of the uses outlined here, nor any limit on the number of leases any person, corporation, or group could hold. The costs of the leases under the terms proposed and the costs of the management of the leased area would, in practice, constitute flexible but significant barriers to large-scale leasing by any lessee.

The basic parts of this illustrative proposal are listed below:

All federal lands now under the management of the Forest Service and the BLM would be open for leasing of the surface. If included in long-term leases of the land surface, the land would then be open for mineral (including coal, oil and gas, and other energy minerals) development by other lessees, with such mineral leases being issued by the lessee of the surface. Since the surface of existing or proposed wilderness areas would be eligible for long-term leases for reservation or conservation purposes, mineral exploration and development would be extended to such areas when leased. Wilderness and other unleased preservation areas would be subject to present laws. Lessees of the surface resources for any purpose would have the power to establish reasonable controls over methods of access and proper site use and restoration by lessees of the mineral interest but would not have the power to refuse mineral development. Royalties might be at 12.5 percent of the value of oil and gas produced and 5 percent of gross output for metals. One-fourth of the royalties would go to the leaseholder for the land surface and the rest to the federal Treasury for distribution to states and counties in accordance with present law. This division of the royalties would provide incentive to the surface lessee to permit mineral development under carefully controlled conditions.

The lessee of the surface could permit the use of the surface for outdoor recreation or could close it to such use; fees or charges for recreation use could be levied and would accrue to the lessee of the surface. Specific conditions about outdoor recreation would not be included in the leases. It is highly probable that most lessees of the surface would tolerate recreation use and would not attempt to collect fees; the experience of the larger forest industry firms on their own land confirms that judgment. Simple public relations dictate making the land available for public use, but the poor income prospects of charging fees, collecting them, and excluding people who do not pay them lead to the judgment that no effort would be made to collect such fees.

Forestlands capable of producing 85 cubic feet or more of wood per acre annually under good natural (but not intensive) forestry would be available for lease by forest industry firms, wilderness reservation groups, or any other interest groups.   However *all* the following conditions would apply:

• The lease period would be 100 years—long enough for two or more complete rotations under intensive forestry, even in the Douglas fir area.

• The lessee would buy and pay immediately for all the standing merchantable timber on the tract, at prices reflecting present stumpage prices for sawtimber and pulpwood.

• The lessee would pay a continuing annual rent for the site and for the immature timber, for the whole period of the lease, at a rate of 5 percent annually on the estimated value of the site and immature timber, adjusted annually in proportion to the Implicit Price Deflator for the gross national product.

• The lessee would be free to harvest standing merchantable timber on a schedule of his own choosing, including never harvesting it if he chose; but the lessee would be bound for the whole period of the lease and would be subject to forfeiture of performance bond if payments for the lease were not kept up.

• The minimum acreage per lease application, under this provision, would be 90,000 acres. In order to get leases of this size in reasonably compact blocks, it might be necessary to allow the inclusion of a small proportion (say, 10 percent) of the land to have a lower timber production capacity. The 90,000-acre figure has been chosen to provide something like a minimum efficient forest management unit; large firms would almost surely seek larger areas or more than one lease.

• The lessee would assume all costs of management of the leased area, including the provision of roads and the control of fires. No specific management requirements, such as reforestation or other timber-management practices, would be imposed in the lease; but the compensation for improvements at the end of the lease would provide strong incentive for maintenance of the forest stand. The costs of such leases would be bearable by the lessee only if efficient and relatively intensive forestry were practiced.

Conservation groups traditionally critical of the forest industry may consider this section of the proposal as being unreasonably generous. However, most of this timber ultimately would be sold to the same or a similar forest industry firm under existing federal land administration. Under my proposal, the lessee would get the timber on leased lands sooner and would more quickly move the forestland into productive new forests. In either case, the firm would pay for the timber at the estimated market price.

Wilderness lovers and other conservationists would have the option under the pullback provision (discussed on page 000) of taking a lease on as much as one-third of the area proposed for any long-term lease by a timber processor, if in their judgment the value of the site for wilderness, recreation, wildlife, or other uses were sufficient that they would be willing and able to pay as much for the timber and site as the timber processor would pay. Since leases of this kind would be open to any lessee prepared to pay the price, wholly new timber-processing firms might be able to enter the business by means of such leases. While established timber processors would have many advantages, they would fall short of a monopoly position on either obtaining such leases or on the prices paid for them. This type of lease would not be available on the extensive areas of "commercial" forests where productivity is below the level of 85 cubic feet of timber per acre annually; timber on such lesser productive sites would be sold, if at all, under present federal land laws.

Forests owned by federal, state, and local governments in the United States customarily have not been leased for long periods, yet this practice is common in other areas, for example, British Columbia and Alberta. In addition, some American forest industry firms have long-term leases for forestland in other countries.

Established public land-using ranchers or associations of such ranchers could lease tracts of grazing land on which they now have grazing permits. The lease term would be fifty years and the minimum area per lease would be 5,000 acres. The annual cash rental per acre would be equal to the U.S. average farm value of 1 pound of beef. The lessee would assume all costs of management, including roads, fencing, water development, and fire control.

Data on beef prices have been collected by the U.S. Department of Agriculture for decades and are readily available. The present formula for establishing grazing fees on national forests and grazing districts includes, as one item, the average farm price of beef, hence the idea of gearing grazing rentals to beef prices is not new or novel. If continuing inflation cheapens the value of the dollar, presumably this leads to higher beef prices and to greater value of the forage from the grazing land. Thus there would be an automatic adjustment in grazing fee lease rates, as inflation or other factors pushed up the price of the beef produced from such land.

A rental of this size economically can be paid only on rangelands producing 1 Animal-Unit-Month (AUM) of forage per 4 acres or less (hence higher output per acre). For land producing 1 AUM from 4 acres, this means a cost more than twice as high per AUM as average current grazing fees on national forests and BLM lands. Ranchers would seek to lease federal lands on these terms only if they were convinced they could make the rangelands more productive or if they feared loss of the land to other user groups. Comparatively few parts of the Forest Service and BLM grazing lands can produce 1 AUM from 4 acres or less. These lessees would have no specific grazing management restrictions in their leases but the lessee would be bound for the full period of the lease, also subject to the performance bond, and subject to a degradation penalty. Under these conditions, no rancher could afford to lease public lands without careful and relatively intensive and continuing conservation management; short-term exploitation would be too costly, as would low-producing rangelands. Because ranchers typically invest less relative to output than do timber processors, and because grass grows on a shorter rotation than trees, the length of the leases here could be shorter than for timber-growing lands. Because the use of federal lands for livestock grazing is intimately interrelated with the use of private lands for livestock production, and because such relationships have developed over many years, it would be inappropriate to allow outsiders to bid for this type of lease. To avoid misuse of this type of lease on lands not intended for this specific land use, the lessee should be forbidden to cut or sell timber from the land. On the other hand, leasing for minerals development would not only be permitted but encouraged.

Private persons, groups, or associations—including conservation, preservation, and similar groups—would be eligible to lease any public lands, for a lease term of fifty years, up to a total of 60 million acres for all such leases. If the lands chosen could qualify for either an annual timber growth of 85 cubic feet or more or a grazing capacity of 1 AUM per 4 acres, the lessee would have to pay the same price as the timber firm or rancher who leased these productive forest- or rangelands. Should the sites not qualify for these timber or grazing leases, leases would require no cash down payment, and the annual rental would be only $1.00 per acre. However, in this case, lessees would be prohibited from harvesting timber. Under both arrangements, the lessee would pay all management costs, including roads, fire, and exclusion of trespassers. As noted, these lessees would receive one-fourth of any mineral royalties. No requirements for management of the area would be imposed in the leases, but these leases might be subject to the same degradation penalties as are grazing leases. The fact that this type of lease would be available to any interested group, plus the provision for leases by units of local government, make it unnecessary to require access to the leased area by the general public. The minimum area covered by leases of this kind would be 7,500 acres.

States and units of local government may select for lease up to 15 percent of the land within their boundaries, or may select for lease all of the federal land if the total federal land is less than 15 percent of the total land area in the jurisdiction.   If the land is forested with a productivity under good natural forestry of 85 cubic feet or more wood annually, state and local governments would have to meet the same terms as other lessees. For all other land, the leases would extend for fifty years on payment of annual rental of $1.00 per acre. The local government would assume all management costs, including roads, fire control, provision of recreation facilities (if any), and so on. Where the area of federal land within the jurisdiction exceeds 10,000 acres, the minimum area for lease would be 10,000 acres. As surface lessees, such units of local government would secure one-fourth of the royalties from mineral development, plus any amounts accruing to them under legislation now applicable to this kind of mineral development. The lessees would be forbidden to sell timber from forests incapable of producing 85 cubic feet of wood per acre annually.

The forgoing illustrative proposal is a package deal, in the sense that every part is essential to the whole. Economically, some parts are defensible only because of other parts; politically, some proposals will be acceptable, only because of other parts.

Industry, government, and many other groups will find the proposal that conservationists and preservationists be allowed to lease as much as 60 million acres for forty years at a nominal rental acceptable only because (1) other leases are available to them, and (2) the leased lands will be open to minerals development. Or, to take another example, the proposal to lease large blocks of the most productive forestland for long periods to forest industry firms will be acceptable only because (1) conservationists and others will have the opportunity to lease large areas of such land or of other land themselves; and (2) the forest industry firms will be paying for the merchantable timber and for the opportunity to grow even more.

Pullback—An Innovative Proposal

"Pullback" is a new idea which could significantly affect future federal land use and management. It is most directly applicable to long-term leasing of federal lands, and the discussion which follows is in terms of its use in federal land leasing, but perhaps the idea could also be applied to the sale of federal lands, if that major alternative were chosen.

The pullback might make increased leasing of federal lands both socially more acceptable and operationally more practical. Under this concept, any person or group could apply, under applicable law, for a lease on a tract of federal land, for any use they chose; but any other person or group would have a limited time between filing of the application and granting of the lease in which to pull back a part of the area applied for. The person applying for the area pulled back would be required to meet the same terms and enter into the same long-term leasing arrangement as the original applicant, but the pullback applicant could choose a different use than that proposed by the original applicant. Thus, competition would arise between uses as well as users.

The right to pullback would be unconditional—that is, available to anyone who could meet the terms, without the consent of the original applicant or of the federal agency. The proportion available for pullback could not be too large, lest it jeopardize the whole lease proposal; yet the user of the pullback authority should have the opportunity to take a significant share of the total—as much as one-third perhaps—if he chose. The one-third limit is admittedly arbitrary, but it seems to meet these criteria. The minimum acreages per lease were established on the assumption that one-third of the area applied for might be pulled back, with the remaining two-thirds meeting the general criteria for each type of lease.

For example, a forest industry firm might apply for a lease on a substantial tract of federal forested land, for a period long enough to harvest present stands of timber and to grow at least one rotation in an economically sized area. A group interested in wilderness, or in developed outdoor recreation, or in wildlife, or in any other use of the land authorized under the law would be permitted to file a pullback application for as much as it chose but not more than one-third of the area applied for by the forest industry firm. The pullback applicant would have to meet the same terms (including the same length of lease) as the original applicant, but otherwise its pullback application would not be subject to refusal by either the original applicant or the federal agency. Some limitations might have to be included in the law, to require that pullback applications be made only for reasonably contiguous lands in order to prevent pockmarking of the whole tract with scattered 40-acre pullbacks. And pulled back areas should not be allowed to further complicate problems of access. There might also have to be some minimum acreage per pullback—no single acre pulled back in the midst of a large lease area; and probably the number of pullbacks per lease should be limited to no more than three.

These same provisions would apply to any pair of uses, in any sequence, or to rival parties who desire the area for the same use. A wilderness lease application would be subject to pullback from recreation groups, for instance; or an application by one forest industry firm for a heavily forested area would be subject to pullback from another such firm.

At present an individual, corporation, or group seeks the use of some tract or area of federal land, for a specific use, under applicable law. If there are others who are interested in the same area, pressure is exerted on the agency that manages the land in order to thwart the original applicant. Those critical of the first applicant use legal means to prevent the lease from being awarded, lobbying Congress and other elected officials. The process is typically an adversarial confrontation, and often no direct bargaining takes place. Mostly there is no solution possible except a victory by one party and a loss by the other. Intermediate or compromise solutions which would more nearly meet the needs of each are almost impossible to achieve.

However, under the pullback procedure, different groups interested in the use of federal lands would be forced to compete and to bargain with one another. The form of bargaining differs from customary competitive bidding in some instances. Every applicant would be vulnerable to the pullback application of every other potential user, and hence would be in competition with the other. This forces the original applicant to consider public or social values, for their neglect would subject him to pullback applied by other groups. And requiring the user of pullback authority to meet the terms of the original applicant largely eliminates deliberately obstructive actions. Competitive bargains among rival interests would be necessary: If you stay out of my area of interest, I will stay out of your area; if you try to impede my use of federal lands, I will retaliate when you try to use federal land.

Use of the pullback provision could lead to collusive action among applicants for federal land. But the competitive bidding process, as it is applied to the federal lands, has not been wholly free of collusive actions, and historically "sweetheart" deals between a federal agency and a private party are not unknown. Collusion in the use of pullback would be as subject to legal action as is collusion in competitive bidding for federal timber or oil. However, it is highly probable that the pullback in operation would lead to less, not more, collusion than now exists, since the pullback would be available to any interested party without consent of the original applicant, the first pullbacker, or the federal agency. To a substantial extent, the pullback provisions would be self-policing.

The pullback provision would also provide a mechanism for guaranteeing that each party adheres to the terms of the negotiation. Bargaining among rival interests in federal land though uncommon is not unknown today. Until now no adequate mechanism for enforcing of agreements has existed, but under the pullback provision, any party to a bargain would certainly face disciplinary or retaliatory action if they failed to live up to the terms of the bargain.

The pullback authority would direct the energies of all claimants and counterclaimants for federal land in a manner closely comparable to the operation of a competitive market. By using the pullback provision, bargaining and competition would replace lobbying in an attempt to influence actions of the federal agencies and administrative action of public agencies would be replaced with bargaining among individuals and groups.

The pullback provision would substantially reduce the use of delaying tactics which are used frequently today. This is a natural enough tactic, given the present administrative and management structures for making decisions about federal land use, but it is also costly to society as a whole. The possibility of disciplinary or retaliatory action by other interest groups would make the rewards of stalling more dubious. Laws establishing the pullback authority might also provide for time limits within which actions had to be taken—for example, a year between filing and granting of the original lease, or six months for the pullback applicant to meet the terms of the original applicant.

Of course, the pullback provision would present operational problems, but they should not be insurmountable. Pullback should be no more onerous to the federal agencies than are the present sale and lease laws. One should never underestimate the difficulties of implementing a new idea, but, at the same time, one should also avoid comparing difficulties under one set of laws with some, imagined frictionless, operation of another set of laws.

The pullback provision would provide substantial help to politically weak contenders for the federal land, to local interests of all kinds, and to special interests as well. Currently those with relatively weak or limited powers on the national or regional scene often suffer total defeat when they wish to influence federal land management or seek to secure the use of some area for their particular needs. By not requiring the consent of either the original applicant or the federal agency, pullback would greatly improve the competitive position of these special interests and at the same time not interpose a serious or fatal barrier to the original applicant.

In the future we likely will experience increasing competition for the use of federal land, which in turn will exert pressures on every user group to consolidate and strengthen its position. Long-term leasing offers one avenue for such improvement, but so does the pullback provision. For instance, the wilderness groups currently place their faith in their political strength, as did the ranchers of a generation ago. A long term lease might provide more assurance for the future than reliance on continual political strife. The use of pullback might be more effective in limiting the inroads of rivals seeking use of federal land than is fighting in the courts, administrative halls, or legislative committee rooms. The various interest groups would do well, I think, to ponder their future security under each of the major alternatives for federal land management, particularly estimating how long-term leasing and pullback could enhance their own interests and security.

Conclusion

Although I believe that major changes in federal land management seem likely in the future, the direction and the specifics of these changes are less clear. The federal lands are too important not to be managed efficiently for their greatest economic output. The inefficiencies in present federal land management seem disturbingly great to many observers. Nonetheless, I reject any idea that we today are less imaginative and resourceful than the men and women who pressed for the establishment of the national forests, national parks, and grazing districts. We too can innovate; let us try.

Notes

1. U.S. Senate, Subcommittee on Public Lands and Reserved Water of the Committee on Energy and Natural Resources, *Workshop on Public Land Acquisition and Alternatives*,

Committee Print, 97th Cong., 1st sess., Publication No. 97–34, October 1981. This paper is an expanded and extended discussion of the paper I presented at the 1982 workshop.

2. Public Land Law Review Commission, *One-Third of the Nation's Land* (Washington, D.C., Government Printing Office, 1970) p. 1.

3. Marion Clawson and Burnell Held, *The Federal Lands: Their Use and Management* (Baltimore, Md., Johns Hopkins University Press for Resources for the Future, 1957) p. 5.

4. Marion Clawson, *The Federal Lands Revisited* (Washington, D.C., Resources for the Future, 1983).

5. An excellent account, which shows the language of the 1974 Act, what was stricken in the 1976 Amendment, and what was added by the latter is found in Dennis C. LeMaster, "The Resources Planning Act As Amended By the National Forest Management Act," *Journal of Forestry* vol. 74, no. 12 (December) 1976. For a good discussion of the 1976 Federal Land Policy and Management Act, see *Arizona Law Review* vol. 21, no. 2 (1979), which consists of a large number of articles specifically focused on this act.

6. John V. Krutilla, and John A. Haigh, "An Integrated Approach to National Forest Management," *Environmental Law*, Lewis and Clark Law School–Northwestern School of Law, vol. 8, no. 2 (Winter) 1978, p. 413.

7. LeMaster, "The Resources Planning Act."

8. *Arizona Law Review* (1979).

9. Marion Clawson, *The Economics of National Forest Management* (Washington, D.C., Resources for the Future, June 1976).

10. Barry S. Tindall, "Fees and Charges in Recreation and Parks—A Review of State and Local Government Actions," a statement presented to the Subcommittee on Public Lands and Reserved Water, Committee on Energy and Natural Resources, U.S. Senate, June 15, 1982.

11. John L. Walker, "National Forest Planning: An Economic Critique"; Douglas R. Leisz, "The Impacts of the RPA/NFMA Process Upon Management and Planning in the Forest Service"; John V. Krutilla, Michael D. Bowes, and Elizabeth A. Wilman, "National Forest System Planning and Management: An Analytical Review and Suggested Approach," in Roger A. Sedjo, ed., *Governmental Interventions, Social Needs, and the Management of U.S. Forests* (Washington, D.C., Resources for the Future, 1983); and Richard W. Behan, "RPA/NFMA—Time to Punt," *Journal of Forestry* (December) 1981. I attended a meeting of the Western Forest Economists at Wemme, Oregon, in May 1981 at which there was a great deal of criticism and no open defense of the Forest Service planning process under these acts. Nelson (see note 30) has pointed to the great costs and generally indecisive results of BLM planning under its acts and in response to court decisions.

12. Richard M. Mollison, "The Sagebrush Rebellion: Its Causes and Effects," *Environmental Comment* (Washington, D.C., Urban Land Institute, June 1981).

13. Ibid.

14. Phillip N. Truluck, ed., *Private Rights and Public Lands* (Washington, D.C., The Heritage Foundation, 1983).

15. Paul W. Gates, *History and Public Land Law Development*, prepared for the Public Land Law Review Commission (Washington, D.C., Government Printing Office, November 1968). See particularly chap. 12 and 13.

16. Denise A. Dragoo, "The Impact of the Federal Land Policy Management Act Upon Statehood Grants and Indemnity Land Selections," *Arizona Law Review* vol. 21, no. 2 (1979), p. 395.

17. Public Land Law Review Commission, *One-Third*, p. 246.

18. Gates, *History* p. 347.

19. John Francis, The Brookings Institution, unpublished manuscript.

20. Gates, *History*, pp. 336–339.

21. Ibid.

22. Ibid.

23. Gary D. Libecap, *Locking Up the Range; Federal Lands Controls and Grazing* (Cambridge, Mass., Ballinger, 1981); John Baden and Richard Stroup, eds., *Bureaucracy vs. Environment: The Environmental Costs of Bureaucratic Governance* (Ann Arbor, University of Michigan Press, 1981); Richard Stroup, "Some Land Use Concepts: Comments," in Roger A. Sedjo, ed., *Governmental Interventions, Social Needs, and the Management of U.S. Forests* (Resources for the Future, Washington, D.C., 1983); Richard Stroup and John Baden, "Property Rights and Natural Resource Management," *Literature of Liberty* vol. 2, no. 4 (1979); and William M. H. Hammett, Richard Stroup, John Baden, and Steven Hanke, "Privatizing Public Lands: The Ecological and Economic Case for Private Ownership of Federal Lands," *Manhattan Report on Economic Policy* vol. 11, no. 3 (May 1982). This is a publication of the Manhattan Institute for Policy Research.

24. Bureau of Land Management, U.S. Department of the Interior, *Public Land Statistics 1980* (Washington, D.C., Government Printing Office, 1980) pp. 3 and 5.

25. U.S. Department of Commerce, Bureau of the Census, *Historical Statistics of the United States, Colonial Times to 1970*, Washington, D.C., House Document No. 93–78, 93rd Cong., 1 sess., September 1975, p. 429.

26. Marion Clawson, *Uncle Sam's Acres* (New York, Dodd, Mead Co., 1951), pp. 86–89.

27. For two of the more convenient sources, see news story in *Federal Lands*, June 21, 1982, a publication of McGraw-Hill, New York, p. 3; and news story in *Outdoor America*, a publication of the Izaak Walton League of America (Spring 1983) p. 6.

28. Marion Clawson and Burnell Held, *The Federal Lands: Their Use and Management* (Baltimore, Md., Johns Hopkins University Press for Resources for the Future, 1957) chap. 6. On page 348*f* we refer to a proposal for a general forest land corporation made by the Forest Products Industry of Oregon, a committee sponsored by Lewis and Clark College and Reed College in Portland, Oregon, in 1956.

29. Ibid., p. 347.

30. Robert H. Nelson, "Making Sense of the Sagebrush Rebellion: A Long Term Strategy for the Public Lands," a paper prepared for the Third Annual Conference of the Association for Public Policy Analysis and Management, Washington, D.C., Oct. 23–25, 1981.

31. Ibid.

32. Marion Clawson, "An Economic Classification of U.S. 'Commercial' Forests," *Journal of Forestry* vol. 79, no. 11 (November 1981).

33. Nelson, "Making Sense."

34. Dennis E. Teeguarden, "A Public Corporation Model for Public Land Management," a paper delivered at a Conference on Politics vs. Policy: The Public Lands Dilemma, Utah State University, Logan, Utah, April 23, 1982.

35. Ibid.

36. Ibid.

11

Sharing Federal Multiple-Use Lands

Historic Lessons and Speculations for the Future

JOHN D. LESHY

Allocating management responsibility for federal lands among federal, state, and local governments and the private sector has been the subject of debate for two centuries. As the Sagebrush Rebellion reminds us, the issue scarcely has been laid to rest. Notably absent, however, has been a discussion of the problems that result from existing arrangements. In this chapter my task is to provide such a focus, paying particular attention to those proposed changes involving less than outright, unrestricted transfer of title to federal lands.

A logical starting point is to examine briefly both the historic evolution and the current allocation of management authority. As every student of the subject knows, both are remarkably varied. History provides numerous examples of different kinds of federal sharing or divestiture of management authority: outright unrestricted grants to states or private interests, and—of more immediate interest—restricted grants in which the federal government retains certain ownership interests or supervisory powers, or has otherwise conditioned transfers in order to limit how these lands subsequently should be managed. Moreover, the federal government has often delegated important management responsibilities to the states, local governments, or the private sector even in areas where it has retained absolute title. Thus, federal land management today is characterized by a marked sharing of power with various nonfederal entities. Altogether, these arrangements comprise a broad array of schemes for disposal and management of federal lands and give insight into special problems they might create.

We would do well to look at this history and these arrangements for guidance. Although a comprehensive evaluation is beyond my energies and abilities, I propose to give a frankly impressionistic review, offering some tentative and perhaps tendentious conclusions about what this record teaches. The broader policy arguments for retention or disposal (whether to nonfederal governments or to the private sector) that are explored elsewhere in this volume are of concern here only by implication.

Because of differences in climate, topography, and historic settlement patterns, the federal government's relationship to state and private interests concerning landholdings has been qualitatively different in the West. Furthermore, current proposals for change focus on federal multiple-use lands, and since the West is where most of these lands are found, the discussion that follows is confined almost exclusively to that region.

Peering into the future is always an uncertain business even where, as here, there is a rich historic record to provide guidance. It is made more uncertain by the fact that, along with the South, the West is the region experiencing the fastest population growth, and, in addition, its population is now the most urban of all regions. Predicting what new constituencies of interest will emerge to overshadow old ones, what new dimensions will be added to old problems, and how the politics and, in turn, the policies of federal land management will be altered requires a measure of guesswork which should temper reaction to the thoughts that follow.

Federal Power Over the Federal Lands

Before addressing proposals for basically altering the existing mix of federal and nonfederal authority over federal multiple-use lands, we should consider how much power the federal government has to effect such alterations. While the constitutional trappings adopted by the Sage-brush Rebels purport to inspire some lingering disagreement about the matter, it seems to me that one principle of constitutional law is now as well settled as any can be; namely, that the federal government's power over federal lands is subject only to relatively minimal constitutional restraints, such as an obligation to provide due process and to protect valid existing property interests against confiscation without compensation. We should, I think, take quite literally the Supreme

Court's ritual pronouncement that the power of Congress over federal lands is without limitations. This means in turn that the important issues concerning management of these lands are resolved almost exclusively through the political process.

At one extreme, the federal government has the power to transfer these federal multiple-use lands outright to the states or to the private sector. Or it could act in a more limited fashion, by transferring part of its title or management powers to the states or the private sector. At the other extreme, the federal government not only could retain ownership of these lands, but also assert exclusive power to manage them— entirely preempting state and local laws and excluding the private sector. Here, too, the federal government could act in a more limited way, by transferring use rights or an ownership interest while restricting uses, even after transfer, in almost any way it sees fit.

These conclusions scarcely are surprising, because history has provided examples of each extreme (and most points on the spectrum in between), all of which have been implemented, at least in this century, without the courts questioning federal authority to effect such arrangements. The existence of such sweeping federal power means that, while constitutional "principles" are excellent fuel for political rhetoric, they do not seriously affect patterns of federal land disposal and management. These patterns instead are limited primarily by more mundane, if no less real, political considerations.

Protecting Secondary Uses During Long-Term Disposition of Federal Lands for Specified Uses

Although one cannot seriously question the federal government's legal power to make and condition transfers of federal land in almost any way it sees fit, one also must consider the practical difficulties in making and enforcing such conditions. This issue is of considerable importance because several current proposals incorporate certain kinds of continuing restrictions. Some so-called privatization proposals call for disposal of title accompanied by restrictive covenants or other legal limitations to ensure that the land, once transferred, will continue to serve other secondary uses in a way which would assumedly not result from the incentives ordinarily provided by private ownership. Perhaps the most common objective would be to foster protection for fish, wildlife, watershed, recreational, and esthetic values on lands disposed of predominantly for timber, range, or mineral exploitation. But that necessarily would not be the exclusive concern; for example, some land, disposed of predominantly for recreational use, could also provide a continuing opportunity for mineral exploitation.

These kinds of restrictions need not be limited to privatization schemes. Other proposals advocate federal retention of title but transfer management to private users under long-term leases or equivalent arrangements which contain similar kinds of restrictions. Still others delegate management responsibility to states under continuing federal restrictions. All such proposals depend heavily upon how easily one can enforce restrictions designed to preserve other uses and values of these lands while they are made subject to what might be described as long-term management for certain dominant uses.

A review of previous federal efforts to achieve such objectives reveals some practical considerations concerning enforcement of such restrictive conditions which, in turn, can inform debate on current proposals.

Patterns of federal disposition and control

One consequence of the dominance of the political process in setting federal land policies is that, like the process itself, the universe of public land disposition and management practices is extremely untidy. The inquisitive student can uncover examples of almost every kind of arrangement involving lands acquired by the United States from foreign governments, ranging from outright unrestricted disposal of ownership to the issuance of permits for sharply limited purposes that are revocable at will by the federal managing agency. This broad spectrum might usefully be divided into three general categories.

The first, and historically most common, category is the outright unrestricted grant (often made, it is true, after compliance with certain threshold conditions such as residence and cultivation in homestead grants, or organization of a policy and adoption of a constitution for statehood grants). With few exceptions, once such land is transferred into nonfederal hands, the only federal restrictions on its use are those which apply generally to all nonfederal lands.

The second category includes those lands in which the United States permanently divests itself of some ownership interest but either imbues the transferred title with restrictions that limit how the new owner is to manage the land or retains some limited ownership interest itself.

The third category embraces transfers of less than a permanent ownership interest, where the United States restricts how the land is managed, as in typical leases, licenses, or permits.

There are, of course, borderline cases. For example, I would classify timber sales and mineral leases in the third category even though, functionally, ownership of the timber or mineral is transferred. On the other hand, I would classify the stock-raising homestead grant of title to the surface of land in the second category, because it involves permanent transfer of a resource where the principal limitation on how it is subsequently managed arises from the federal government's retention of mineral rights. Occasionally these categories overlap; for example, outright transfers of title to ranchers under homestead laws (falling either in the first or second category depending upon whether the United States retained mineral rights) usually furnish the requisite "base property," justifying the issuance of a permit entitling the same rancher to graze livestock on nearby federal multiple-use lands (the third category). I trust, however, that these categories are distinct enough to be useful for my limited purpose here.

Restrictions on use after outright unrestricted transfer

Experience with the first category illustrates that even full unconditional transfers of federal land do not place that land beyond federal regulatory control. Taking just one example, coal no longer may be mined from a tract of land in Illinois which passed from federal ownership more than a century ago except in compliance with the Surface Mining Control and Reclamation Act of 1977.[1] In other words, federal land which becomes private under some transfer scheme derives no special protection from subsequent federal regulation even though the transfer was unconditional. In legal terms, the role of the United States government as sovereign differs from its role as proprietor or landowner.

The continued existence of sovereign power to regulate uses of purely private land suggests that there need be no real concern over removing multiple-use lands from federal ownership. If it turns out, for example, that wildlife or esthetic values are downgraded when a tract of national forest land is sold to a timber company, then the federal government need only enact legislation to provide protection for such values on those lands. The fact that restrictive federal regulation comes after unrestricted transfer should not limit its effectiveness.

There is a kernel of truth in this argument, but there are other problems to be considered. First, there is the obvious practical difficulty of getting Congress to regulate after transfer. Second, Congress may lack the political will to protect wildlife or esthetic values on all private land to the same extent it finds desirable to do so on recently transferred land. And to the extent that the rationale for limiting regulation is based primarily on the length of time the regulated lands have been in private hands, congressional regulation may be legally impermissible on the same basis that the courts have struck down so-called spot zoning ordinances, which treat similarly situated parcels of land differently.

Further, the federal government's power to regulate in its sovereign capacity may be somewhat weaker than its power to regulate as proprietor. Although the latter is essentially without limitations, the former is restricted, at least formally, by such constitutional requirements as assuring a connection to interstate commerce and, under the Tenth Amendment, the integrity of state governments.

In short, it is not easy for Congress to enact legislation requiring special protection for values on lands previously disposed of without restriction. On the other hand, Congress has had no difficulty in imposing restrictions in the grants themselves.

Another, simpler way to restore such protections after divestiture would be for the federal government to reacquire the lands previously transferred. History offers us several examples of federal acquisition or reacquisition of lands in order to protect values or to serve purposes which private owners were unable to undertake. The most notable acquisition programs for multiple-use lands have involved the eastern national forests (mostly previously logged, mountainous forestlands) under the Weeks Act of 1911,[2] and what are now the national grasslands (mostly failed homesteads in the Dust Bowl) pursuant to New Deal legislation designed to "correct maladjustments in land use."[3] Of course, reacquisition is not a happy alternative, because it suggests that outright privatization has failed to provide the benefits it was designed to give.

In sum, although it is useful to remember that the federal government has relatively broad powers to regulate private land, both politics and law may limit this power, especially where the objective is to protect former federal lands selectively. Thus, if federal restrictions on the use of lands now managed by the federal government for multiple use are desirable, they most effectively should be formulated and applied at the time of transfer.

Restricted transfers of ownership

In this category I include the transfer of millions of acres of federal land to states upon statehood, at which time the federal government required that the land be managed to produce maximum income for common schools in the state. Dissatisfied with what it felt was mismanagement of federally granted lands by the older states, in later statehood acts Congress included increasingly severe restrictions, such as requiring that all dispositions of such lands be made at public auction for not less than the appraised market value. The remedy for breaching such requirements often was specified as judicial enforcement upon suit by the United States.[4]

Another such example is found in the Recreation and Public Purposes Act,[5] under which title to federal lands can be transferred to state and local governments for certain purposes. These grants by law contain reverter clauses, allowing the federal government to recover the lands if they are no longer used for the purposes for which the grant was originally made. Special grants have been made in which Congress included continuing restrictions on the use of the granted property. One of the best known of these is the controversial Hetch Hetchy grant in Yosemite National Park to the city of San Francisco.[6]

Numerous nineteenth-century land grants to railroads required the recipients to sell at controlled prices a large part of these lands in small parcels to *bona fide* settlers.[7] Described as a "sop to the homestead element,"[8] they reflected an apparent desire on the part of Congress to further westward expansion by yeoman farmers.

Another type of transfer belonging in this category is the disposal of the surface rights and retention of the mineral estate, or vice versa. The best-known of these grants occurred under the Stock-Raising Homestead Act, under which many millions of acres of federal land surface were transferred to ranchers between 1916 and 1977, with the federal government retaining the mineral interests.[9] The Stock-Raising Homestead Act explicitly makes the federally reserved mineral estate dominant, so that the homesteaders who took title to the surface had no right to prevent exploitation of the reserved minerals. The only compensation Congress provided to the surface landowner was for damages to crops and permanent improvements, and not to the land itself or to any other feature of the surface owner's operation.

From my knowledge of these grants, I have formed some general impressions, which allow me to draw some tentative lessons. First, concerning the statehood grants, a number of studies support the conclusion that the command to obtain fair market value in the leasing or selling of these lands has been widely ignored and that, overall, these trust restrictions have been honored more in their breach than in their observance. Although these grants usually have authorized judicial enforcement, court review has been rare. Noting the apparently widespread violations of these restrictions, the Public Land Law Review Commission (PLLRC) recommended that they be repealed.[10]

Other special restrictive grants have suffered a similar fate. When Congress—finally ending a lengthy battle in 1917—granted San Francisco the right to extract water and power from the Yosemite National Park by damming and inundating a valley whose beauty rivaled neighboring Yosemite Valley, Congress forbade the city from selling the power so generated for resale. That prohibition has been openly violated from the very beginning, even though the United States eventually sought judicial enforcement and a landmark 1940 Supreme Court decision upheld the restriction. More recent litigation has been similarly ineffective, even though the courts continue to acknowledge that the intent of Congress has been thwarted.

In one celebrated instance arising shortly after the turn of the century, the United States sought to enforce a proviso of one railroad grant, charging that the railroad grantee had retained most of the granted land and sold the rest of it in violation of the statutory size and price limits. After the Supreme Court agreed, Congress enacted legislation which revested title to the unsold lands in the United States but which, significantly, ignored those lands the railroad had previously sold to third parties in quantities or at prices exceeding the terms of the grants. Moreover, in taking back title to the unsold lands, the United States paid the railroad grantee the price the latter would have received if it had complied with the restriction by selling to settlers. Thus, even in this exceptional situation, which gave birth to the so-called O&C lands in western Oregon, purchasers from the railroad were fully protected despite their lack of *bona fides*, and the railroad suffered no actual penalty from its breach.

Apart from this almost unique case, the restrictions in the railroad grants turned out to be largely unenforceable and ineffective. A basic problem was in the language of the grants: for example, the restrictions were not clearly labeled as covenants or conditions (a legally significant distinction), failed to set out a clear mechanism for transferring the land to settlers, and were silent both on remedies for violation and on their enforceability by the courts at the initiative of either the executive branch or potential settlers. These ambiguities eventually resulted in a series of Supreme Court decisions—such as one allowing railroads to shield themselves from the duty to sell the lands simply by mortgaging them— which largely negated the restrictions. After years of wrangling, Congress and the executive branch together finally washed their hands of the matter on the eve of World War II.

This is not to say that congressional attempts to include these restrictions were totally frustrated; indeed, some of the land subject to the restrictions passed from railroad ownership, and small parcels actually were sold cheaply to those *bona fide* settlers Congress apparently intended to benefit. But even today railroads own large acreages that had been subject to these restrictions, and undeniably disposed of some of their grant lands in large tracts, at higher than statutory prices, to other than actual settlers, all in contradiction of the ostensible purpose of Congress. The fact that some of the granted lands actually ended up in the hands of the intended secondary beneficiaries seems more coincidental than not.

Separating the surface rights from the mineral estate in the Stock-Raising Homestead grants posed no serious problems for several decades, since what interest there was in development of the federally reserved mineral interest was sporadic. Where mining enterprises desired to develop the mineral estate, they usually negotiated compensation agreements or purchased the surface rights from the owners. To avoid controversy, these settlements often were more generous than the mining companies were legally obligated to offer.[11]

In the 1960s the surge of interest in coal development in the Northern Great Plains (where a sizable fraction of Stock-Raising Homestead grants are found) caused the ranchers as a group to feel that not only their lands, but also their way of life was being threatened as they faced the transformation of the region's economic and cultural base from ranching to energy development. As a result, Congress was persuaded essentially to reverse its decision made six decades earlier. The celebrated "surface owner consent" provision of the Surface Mining Control and Reclamation Act of 1977 effectively made the private surface estate dominant and the federal coal interest subservient.[12]

Another example of the chilling effect that separation of surface and mineral rights can have on development is the experience with the Small Tract Act and other federal laws authorizing disposal of federal surface for recreational purposes.[13] Although Congress reserved the minerals and authorized their development under regulations to be promulgated by the secretary of the interior, successive secretaries have refused to issue such regulations and congressional acquiescence has effectively thwarted the avowed purpose of the mineral reservations. To the chagrin of the mining industry, then, Congress and the executive branch have discovered that, no matter what was intended at the time, the severance of the surface from the minerals restricts or thwarts minerals development.

So far I have discussed situations involving federal divestiture of some type of ownership interest. With respect to water and wildlife (including fish), it is less appropriate to speak of "ownership." As a practical matter, however, Congress has, with some exceptions, generally transferred control over perhaps the most important single resource found on the western federal lands—water—and allowed the states to allocate and manage it. This severance of water from the federal lands has had an enduring effect on shaping the use of all land in the West, both federal and nonfederal (see below).

In essentially the same way, Congress has transferred management of game fish and wildlife species to the states—particularly the power to set seasons and issue licenses for hunting and fishing. However, here too there are significant exceptions; for example, Indian rights, certain anadromous fish, migratory birds, endangered species, and wild horses and burros. Moreover, federal multiple-use agencies continue to manage game and fish habitat and include fish and wildlife in their land and related resource planning processes.

Nonetheless, the states have developed substantial bureaucracies to discharge these responsibilities and also control to a large extent the taking of fish and game on federal multiple-use lands. This control has been very zealously guarded by the states, although recent Supreme Court decisions have made it quite plain that the states have no constitutional protection from a diminution of their role by congressional action.

Congress has maintained its historic deference, although in recent years there has been some erosion in the breadth of state control. Even where strong pressure exists to assert the national interest over that of the state, Congress has acted only with considerable difficulty. In the Federal Land Policy and Management Act,[14] for example, Congress attempted to redefine the respective limits on state and federal power over wildlife on the public lands, but the redefinition became so controversial that it threatened to scuttle the entire act, and the compromise which eventually emerged was, at best, muddled and, at worst, incomprehensible. Shortly after the act passed, the state of Alaska mounted a program to exterminate wolves on large tracts of federal multiple-use land. Although it possessed the power to stop the controversial program, the Department of the Interior refused to do so, and subsequent litigation brought by wildlife advocates to halt the extermination did little to clarify matters. On balance, the controversy underscored the fact that the historic deference to state programs makes it very difficult for a federal agency to thwart a state's will. Extermination of wolves over a relatively large area of federal lands in Alaska was not threatening enough to justify federal interference, even by a relatively activist, environmentally sensitive administration.

Restrictive transfer of less than an ownership interest

Such limited transfers are of course the most typical means by which nonfederal interests use federal multiple-use lands today. This category embraces a multiplicity of arrangements—written or oral leases, licenses, permits, and rights-of-way—for a multiplicity of purposes, recreation, mining, grazing, and timber harvesting being the most important. The legal rights conveyed to the users vary from none to substantial property interests, the tenure ranges from less than a day to many years, and the restrictions on use vary widely. With minor exceptions, however, Congress has given the executive branch substantial discretion in deciding whether and on what conditions to authorize uses of these federal multiple-use lands.

Where Congress has given the executive branch wide discretion to permit and regulate uses of these lands, the historical record clearly shows that, in most cases, federal agency administration of this discretionary power in effect has amounted to turning development or use of these resources over to the private sector on demand. This is especially true for mining, grazing, and many forms of recreational activities, and it has occurred even when the applicable laws did not explicitly require or seemed to prevent such a result.

Mining. The lengthy battle leading up to enactment of the Mineral Leasing Act of 1920 focused on whether the federal government would lease, rather than sell, land containing mineral deposits, and concomitantly, whether the federal agency could refuse to lease the lands.[13] Although Congress answered both questions in the affirmative, mineral leases on most federal multiple-use lands have generally been available virtually for the asking. Today well over 100 million acres are under oil and gas lease, and new leases are issued almost automatically as the old ones expire. Even today little attempt is made to determine consciously, in advance, whether other resource values might outweigh the potential mineral value. The same approach historically has been applied to all other leasable minerals. Although litigation, legislation, and administrative changes during the 1970s resulted in substantial changes in the federal coal-leasing program, the Reagan administration has recently moved to modify the Carter administration's coal-leasing program by deemphasizing the significance of federal coal production targets, and by deferring more to industry wishes in leasing.

This discussion is not necessarily designed to condemn these practices on the part of the executive branch. For one thing, it comports with that body of opinion which argues that the private sector, driven by the competitive market, does a better job than the government in determining the rate of minerals development. For another, the executive branch occasionally has exercised its prerogative to say no to minerals leasing by refusing offers to lease or, as in hardrock mining, by formally withdrawing a particular area from mineral activity. Finally, the leasing terms give the federal government substantial control over whether and how development occurs in order to protect other values. For example, this authority is now sometimes exercised to require special care in protecting areas of special environmental sensitivity. But, on balance, it seems fair to conclude that administration of the mineral leases has not seriously displaced private competitive incentives in determining whether and how mineral development occurs, at least on federal multiple-use lands which have not been withdrawn from consideration. (Given that mineral development is the most lucrative among all uses of the federal lands, the implications of adjusting current arrangements will be more thoroughly explored below.)

Grazing.   The administration of the forage resource on federal multiple-use lands generally follows the pattern of mineral leasing. Although the laissez-faire policy was officially ended by Congress in 1897 on national forests and in 1934 on public domain lands, the preexisting use patterns remained largely, though not completely, undisturbed. Official abandonment of completely unrestricted grazing led to modest fees, some modest regulation, and a formal mechanism for reducing or canceling grazing privileges. In return, these laws offered most graziers certain advantages, of which the most important was perhaps the stabilization of the range which resulted from applying the statutory preference system of the Taylor Grazing Act in such a way as to protect fixed ranching operations, to the detriment of nomadic livestock operators, who lacked privately owned "base property."[16] The system also helped justify increasing federal financial assistance, in the form of range-improvement programs, to that segment of the livestock industry dependent on federal lands.

Congress has generally refrained from creating any legally enforceable property rights in grazing permittees, and concomitantly has given federal agencies sweeping authority to reduce or revoke grazing privileges where necessary to protect or serve other uses. Yet grazing continues to be permitted nearly everywhere there is forage on multiple-use lands. While regulation of grazing permittees seems to be increasing in some respects, and reductions in grazing allotments have periodically taken place in the past few decades, overall there has been no dramatic alteration in the control exercised by private graziers. Repeal of the laissez-faire system scarcely meant, in other words, wholesale replacement of private with public decision making.

Timber harvesting.   The other most widespread and significant commodity use of the federal multiple-use lands—timber production—has historically been managed differently. Federal timber production did not become a major factor in national wood-fiber supply until rather recently. Around the turn of the century, it was made subject to explicit federal control on the most important federal timberlands, the national forests. The volume, tenure, and other terms of federal timber sales have varied widely from place to place and over time, but the multiple-use agencies typically have retained significant control over the rate and method of harvesting, road construction, and other aspects of the sale. Moreover, the system has been administered in such a way as to lead to more vigorous private competition for timber than for range and minerals. Experience with long-term sales, however, such as the 1968 sale of several billion board-feet of timber over a fifty-year period in the Tongass National Forest in Alaska, has not been altogether happy. That sale was canceled after substantial opposition from environmentalists led to lengthy, but ultimately inconclusive, litigation.

Recreation. Most forms of relatively simple private recreation, such as hiking, camping, photography, bird-watching, and rock-hounding, are permitted on most federal multiple-use lands without much regulation. (As noted earlier, hunting and fishing are subject mostly to state control.) Compared to mining, grazing, or logging, such activities usually involve a less consumptive use of, or potential for damage to, federal resources. As a result, written authorization for use is rarely required, and such users are considered casual, implied licensees or invitees. More resource-threatening forms of recreation such as off-road vehicle (ORV) sports went unregulated until the early 1970s, and although President Nixon issued an Executive Order calling for zoning the multiple-use lands to regulate ORV use, its implementation by the agencies has been criticized as being timid and mostly aimed at maintaining the status quo.[17]

Other purposes. Finally, the federal government has, by means of leases or permits with relatively long tenure, transferred areas of multiple-use lands to the private sector for a variety of commercial purposes other than mining, grazing, or logging. These embrace such diverse enterprises as rights-of-way for pipelines and transmission lines, special use permits for ski-areas, marinas, and other recreational developments, as well as for other miscellaneous purposes.

It is difficult to generalize about the experience in this area because of the myriad of purposes served by an almost equal variety of arrangements. Thus some of the lands covered by such arrangements are almost wholly devoted to the narrow purposes for which are leased (for example, ski areas), while others involve uses which do not substantially interfere with most other uses (for example, rights-of-way for pipelines or transmission lines).

Usually the amount of federal land involved is small, as is the impact on the land. Such developments spark concern or controversy only where strong competing interest exists. Most often this competing interest seems to be, as in the celebrated Mineral King ski area proposal, preservation for more primitive forms of recreation or wilderness.

The federal agencies typically retain a large measure of control over such developments in the applicable leases or permits. Intensive recreational developments such as ski areas are treated as natural monopolies whose terms of service to the public are regulated much like public utilities. The length and certainty of lease tenure, however, raise some interesting issues. The user's need for investment security would suggest a need for legally protected tenure, because the facilities authorized by many of these leases or permits require large capital outlays with long pay-back periods.

Yet, several cases illustrate that legal protection for tenure is not always essential. Most of the major alpine ski areas in the West, for example, are located at least partly on national forest land. Applicable law limits the length of permits for such uses to thirty years and, more important, limits the total area to not more than 80 acres. Large modern ski areas need more land than that, however, and the solution applied by the Forest Service, and acquiesced to by the operators, is to issue special-use permits, legally revocable at will, for those permanent facilities constructed outside the 80-acre base permit.[18] Though the operators have no legally protected tenure on the additional land, they have little difficulty in attracting investment capital. Neither the ski operators nor the Forest Service have found it necessary, or even desirable, to ask Congress to enlarge the permissible acreage for tenured permits. I suspect that the Forest Service desires the leverage the revocable permit gives it over the operator, while the operators are not willing to risk disturbance of the status quo by raising the issue in Congress, where outcomes are often unpredictable.

Perhaps the most dramatic illustration of the limited relevance of legal tenure in using federal lands was provided by the initial right-of-way issued for the trans-Alaska oil pipeline. Until struck down by the federal courts as inconsistent with applicable federal law, the federal government was prepared to issue an untenured permit, revocable at will, for permanent facilities vital to the operation of the $8 billion pipeline. The legality of the permit aside, what is particularly interesting is that the companies building and operating the pipeline were willing to invest a large sum on the supposition that the federal government would not do what it had a perfect and almost unconstrained legal right to do—that is, to revoke the permit and shut down the line.[19]

Of course, the irrelevance of legal rights in the trans-Alaska pipeline right-of-way to the companies' willingness to build the facility may not be widely applicable to other federal land use arrangements. For one thing, the companies built the facility with their own funds, and thus were not directly dependent upon outside investors who might have exhibited more nervousness over the legal insecurity. Second, the sheer size of the project and its national security overtones lent substantial support to the belief that revocation of the permit was, as a practical matter, unthinkable. Nevertheless, the ski area and pipeline examples underscore the fact that legal protection for tenure is only one of many factors which influence private investment decisions in federal lands.

Enhancing the Effectiveness of Federal Restrictions

This overview prompts several observations pertinent to a contemporary rethinking of the federal multiple-use lands. First, restrictions imposed with a transfer of ownership are likely to prove ineffective unless the restrictions are framed with great care and certain conditions favorable to enforcement are present. This statement may seem too skeptical; after all, such restrictions are routinely demonstrated in the transfer of ownership of private land, through restrictive covenants and a variety of other arrangements. Why should their efficacy be doubted because a public agency imposes them? The difference, which is critical to federal lands policymaking in general, is that formulation and enforcement of these restrictions by public agencies is much more bound up with a larger political process, and that process is heavily weighted toward preservation of the status quo. This, in turn, means that the more permanent the transfer, the greater the recipient's expectation of substantial noninterference, and the more difficult clear formulation and firm enforcement by the transferring agency becomes. Thus, for example, restrictions on the statehood and railroad grants proved largely ineffective because the transfer of ownership eventually vitiated the federal establishment's political will to enforce the conditions of the transfer.

This first lesson might suggest that federal restrictions should be more effective the less ownership or tenure a nonfederal user of federal lands has but, paradoxically, this is not necessarily true. A second lesson to be drawn from history, then, is that historic expectations and actual uses often (perhaps even usually) outweigh formal characterizations of the rights of federal land users. The status quo toward which the political process is weighted, in other words, is the status quo in fact more than in the law.

A stark example of this is provided by the Stock-Raising Homestead grants, under which the federal government encouraged the development of a ranching economy based on private surface ownership and the use of nearby federal grazing lands. This pattern of use proved so difficult to dislodge that even though the federal government continued to possess the unquestioned legal right to override ranchers' wishes by developing the federally reserved minerals, it eventually found that it lacked the political will to do so. Instead, it chose to transfer a substantial part of the authority over the timing and location of federal coal development from the federal government and private industry to the surface landowners.

The presence of other coal deposits on federal lands where the surface was not privately owned obviously was a factor in the congressional decision; nevertheless, it was made when oil imports were large, the "energy crisis" was a dominant political issue, and coal (especially federal coal) was touted as its best short-run solution.

In retrospect, it was not patently unreasonable for Congress to decide that the ranching subculture was worth preserving, even at the price of diminishing the value of the retained federal mineral interest. However, the point remains that in this instance protection of the *de facto* rather than *de jure* status quo proved to be politically irresistible.

The tendency to protect actual patterns of use rather than the legal rights to use federal lands—reflected in the folk wisdom that possession is nine-tenths of the law—has operated in other areas as well. Early miners on western lands were technically trespassers, yet their lack of tenure obviously failed to halt private investment in their ventures. More recently, existing patterns of grazing and mining were largely maintained on federal lands even after the laws were changed to give the federal agencies greater management authority. On those relatively rare occasions where an agency has moved more aggressively to implement this authority, such as by raising grazing fees toward market value or by reducing grazing allocations where overgrazing exists, Congress has often intervened to modify the agency's legal authority in order to protect the status quo, by, for example, limiting increases in grazing fees or making cutbacks in grazing levels more difficult.

This is not, I might add, limited to the more traditional consumptive uses of the federal lands. A controversy has recently flared concerning minerals development in untrammeled areas of the federal lands. In the Wilderness Act,[20] Congress decided against legally banning mineral ac-

SHARING FEDERAL MULTIPLE-USE LANDS

tivity in such areas, even when it would be inconsistent with wilderness preservation. In the nearly two decades that followed, however, these areas have seen little exploitation. Many factors might explain this phenomenon—perhaps mineral values are low or, given the remoteness of most of these areas, the costs of extraction and transportation are too high. But I would argue that the overriding reason has been the power of expectations created by existing uses (or nonuses) of these lands. Government does not change the status quo easily; in fact, it often serves to retard rather than to hasten change, and many in the minerals industry apparently have calculated that the status quo is not worth changing, because industry investments could be more productively made elsewhere, with less negative publicity. The wisdom of such a calculation has, I think, been demonstrated by recent events—the emergency withdrawal of wilderness areas from oil and gas leasing, the opposition by many in Congress, including some key western Republicans, to such leasing, and a substantial backtracking by the Reagan Administration.

Even the courts sometimes find persuasive the siren call to protect existing uses and practices and yield to it even when the legal principle which ordinarily would be applied is at odds with that result. One need not be cynical about judicial behavior in order to recognize that courts do occasionally strain to avoid decisions which have the effect of altering the status quo, and a variety of rationales and technical legal doctrines are available to justify such results.

These illustrations are useful reminders that long-standing uses of federal lands, even where they have become controversial, have a substantial inertia which is politically difficult to alter, whether these uses are legally protected from change or not. If that inertia can be overcome in order to change the management of federal lands in significant ways, the new course—whether it be privatization, transfers to states, or long-term leasing—quickly will acquire an inertia of its own. In this context, it is worth remembering that the principal purpose for transferring authority over federal lands from the federal government is to lessen, not to maintain or to increase, federal regulation. This deregulation objective contains within itself an inertia which will tend to vitiate any lingering federal regulation. Even the most carefully designed web of legal protections for certain interests, woven into a scheme to change the management and uses of federal lands in fundamental ways, can become irrelevant. Such legal conditions, in other words, merely fix at a specific point in time the peripheral aspects of a basic shift in power, but as time passes larger political forces attending that basic shift ultimately will determine the effectiveness of these lingering, peripheral restrictions.

The process by which such restrictions would be formulated is far more complex than a simple bargaining between two parties over contract terms. Instead, it is typically more public, involves a wider array of interests and, most important, is driven more by the short-term appearance than the longer-term reality of what is being done. For example, shortcomings in the language of the railroad grant restrictions played a large part in their vitiation. It could be argued that such defects were technical flaws, oversights which could be corrected in new grants. But another explanation is more likely—that the failure to eliminate the ambiguities (or to provide an adequate administrative mechanism for federal supervision, as in the statehood grants) was not the result of poor judgment but rather represented a deliberate political choice, a conscious and successful attempt to reduce the restrictions from iron-clad conditions to merely hortatory declarations.

Restrictions are more likely to be effective if they have the support of organized interest groups outside the government, and if these groups have access to the courts. The intended beneficiaries of the railroad grant restrictions—potential settlers—were a diffused, largely unorganized lot, and they themselves were not given any enforcement role. A reasonably persuasive case can be made that the potential beneficiaries of the kinds of restrictions being contemplated today to accompany the divestiture of federal power over federal multiple-use lands are already well organized, and would, as a matter of course, have access to the needed information and to the courts for enforcement purposes. Indeed, three of the most significant changes in federal land management in the last decade have been the rise of organized constituencies for recreation, preservation, and wildlife protection; the passage of the Freedom of Information Act; and the expansion of judicial review.

Were Congress now to turn large tracts of federal multiple-use lands over to livestock, mining, recreation, or timber interests, with restrictions akin to those placed on earlier transfers, the outcome might be different. Nevertheless, the rich and not altogether happy history of restrictive disposals of federal lands is a potent reminder that such restrictions can prove illusory, failing in their mission to safeguard the interests they are designed to protect.

While I have suggested some factors that influence the efficacy of formulating and enforcing protective restrictions, a much more thorough examination should be done if the drive to reevaluate the federal lands gains momentum.

First, a more careful examination could be made of long-term commercial leases of federal lands. The BLM, for example, has issued long-term commercial leases on lands along the Colorado River that have been withdrawn for reclamation purposes. The National Park Service and other agencies routinely enter long-term concession contracts, and since the early 1920s the Federal Power Commission (now the Federal Energy Regulatory Commission) has issued long-term licenses for hydropower generation. It would be fruitful to compare these experiences with the ones I have touched on here, with certain other special forms of federal–private cooperative management (such as cooperative units on national forests under the Sustained Yield Forest Management Act of 1944,[21] and the Forest Service's reliance on private cooperative grazing associations on the national grasslands), with the experience of foreign governments who have entered long-term resource extraction or management agreements with private companies (including American companies who also do business on the federal lands), and with purely private leases.

One aspect that deserves particular scrutiny is the degree of supervision over and enforcement of the terms of the federal lease or contract. Robinson's 1975 study of the Forest Service is somewhat critical of that agency's enforcement of timber sale contract terms.[22] Another more recent examination of public land agencies suggests that while increased advocacy by environmental groups has not changed the allocation of uses on the public lands, it has increased the agency's scrutiny of authorized activities and its enforcement of standard contract provisions.[23] An inquiry into the factors affecting agency behavior and an evaluation of types of restrictions and enforcement mechanisms now in existence, such as performance standards, the use of financial incentives, performance bonds, penalties, arbitration, and so forth, could likewise prove instructive.

Another issue worth pursuing is the question of how many and what kinds of protective restrictions are necessary if federal multiple-use lands were to be disposed of by sale or long-term lease. Expressed another way, what are the shortcomings, from a public multiple-use standpoint, of private management of large tracts of privately owned land? There are several large private landholdings in various parts of the United States—many of a size comparable to a typical national forest or BLM planning unit—that are managed totally or in part for resource outputs, or primarily for such conservation purposes as wildlife protection. Perhaps we do not know as much as we should in terms of management inputs, resource outputs, environmental protection, and recreational access about how management of these lands differs from that of the federal multiple-use lands.

Finally, considering that an oft-touted virtue of our federal system is the availability of the states as laboratories for socioeconomic policy experiments, we should take advantage of the fact that the western states received millions of acres of federal land as statehood settlements, which they subsequently managed or disposed of by a variety of schemes. While most of these grants were in scattered smaller parcels, some were large, and others were combined into larger parcels comparable to federal tracts. A few comparative studies exist,[24] but it remains to be determined if a basis exists for making policy judgments such as whether Nevada or Arizona is better off, the former having disposed of its statehood land grant and the latter having retained nearly all of its grant.

Some Special Implications of Major Changes in Tenure

Mineral resources

From the first attempts to devise a policy for developing minerals found on federal lands, there has been concern regarding the relationship between mining and other uses of the land.[25] Today, well over 100 million acres of federal multiple-use lands are subject to mining claims or mineral leases, with only occasional protests from surface users. Most of this acreage is leased for oil and gas, which by and large involves a relatively small impact on most surface claimants such as ranchers, loggers, and recreationists, at least when compared to other minerals activity. Development of bedded minerals such as coal or hardrock minerals such as gold or copper can involve long-term disruption and even permanent destruction of the surface estate. Yet such mining activities disturb at most only a few thousand acres per year, so the total impact is not great. This is especially evident in comparison to, say, the acreage of federal lands devoted mostly to rangeland, recreation, or timber production. Furthermore, known areas of potential minerals can be excluded from consideration for intensive management of the surface resources by long-term leasing or disposal, and this would further reduce the potential for conflict. Thus, there are ample grounds for arguing that long-term leasing or other disposal of the surface of federal multiple-use lands will not seriously affect minerals exploitation.

Nonetheless, the very factors that supposedly operate to allow the long-term lessee or private owner to achieve better management with greater efficiency could lead also to a greater desire on the part of the surface owner to avoid or to mitigate the disruptions caused by minerals development. On the other hand, it could encourage the surface claimant to engage directly in it. Similarly, the prospect of easier mining access or other profits might lead mining companies to bid on surface rights and, if successful, to enter the ranching or timber business, thereby deriving whatever advantage might be possible from unified operations. Either of these possibilities threatens the status quo, and the prospect of greater competition among surface users and mineral developers for federal resources is not likely, without some adjustments, to be embraced wholeheartedly by either of them.

To the extent that surface users desire to avoid mineral development, they could achieve their purpose under existing law simply by pressuring the federal government to refuse a lease on the mineral estate, or to withdraw it from operation of the Mining Law of 1872.[26] Because that law gives the federal government comparatively little control over the location, rate, and manner of hardrock mining, however, it might be reasonable to expect it to advocate its repeal and replacement by a leasing system.

This thought bears some elaboration. It is seductively easy to assume that current laws governing federal minerals development—particularly the division between location and patent of hardrock minerals under the Mining Law of 1872 and the leasing of most energy and some other minerals under the Mineral Leasing Act of 1920 and similar laws—would continue to apply under any major new scheme emerging from a re-evaluation of the federal lands. But substantial support already exists for replacing the Mining Law with a leasing system, and a major restructuring of federal land management might well provide the occasion for the success of such an initiative.

When the federal government came to address minerals development on acquired federal lands shortly after World War II, for example, continuation of the historic division between location and leasing was rejected and an all-leasing system applying to hardrock, as well as already leasable, minerals was substituted.[27] Furthermore, as Marion Clawson's proposal suggests (see page 224), if the surface lessee is compensated by a fixed share of the federal royalty derived from minerals development, such a royalty would have to be created for hardrock minerals because none now exists. It takes little imagination to see this as a wedge to pry open the larger issue of replacing the Mining Law with a leasing system. A major change in federal multiple-use land management might, in other words, provide an adequate excuse for leasing advocates to achieve the reform that has eluded them for decades. Concomitantly, such proposals are apt to discomfit the hardrock mining industry, and heavily influence its attitude toward such a proposal.

Another way to safeguard the interest of the surface owner or lessee would be to give him or her the right to veto leasing or location of mining claims. This would reverse the usual assumption that the federal mineral estate is legally dominant—that is, the holder of the surface estate has no right to block development of the mineral interest. In 1977 Congress, in effect, eliminated the dominance of the federal coal estate underlying the private surface acquired under the Stock-Raising Homestead Act. That provision requiring the surface-owner's consent was, in a larger sense, simply the latest manifestation of a largely uninterrupted historic trend to reduce, restrict, or otherwise control mining activities in order to protect other uses on federal multiple-use lands, and to give power to do so to the surface user or manager. The Acquired Lands Mineral Leasing Act (1947) gave the surface managing agency (principally the Forest Service) a veto in all mineral leasing decisions on acquired lands. Even on ordinary national forest lands, where the Forest Service lacks a general veto power over mineral leasing, in practice it has a very substantial influence over whether leases are issued at all, and over lease terms and conditions. Similarly, both the Forest Service and the BLM have begun regulating activities on mining claims in order to protect other resource values.

Keenly aware of that trend, the mineral industry would undoubtedly want explicit assurances that any proposals to alter patterns of federal land management would not further restrict access to minerals. At the same time, nonfederal surface owners or lessees may continue to demand a larger role in the process of determining mineral access. It is obvious that some accommodation will be necessary.

Another problem concerns access across nonfederal lands to federal minerals beneath the federal surface. Traditionally, the federal government has played no role in securing such access. The access rights of federal lessees or claimants are controlled entirely by state laws. These laws vary somewhat but, significantly, many states do not provide the miner with the power to condemn access. In such states, access rights must be purchased from a willing seller. To the extent that so-called privatization occurs, these problems will increase. Congress could, of course, ensure development by giving the lessee or claimant the power of condemnation, or by reserving an access right in transfers of land to nonfederal interests. This would require a sharp break with historic practice and it is likely it will be resisted. Here, too, some accommodation will likely be necessary.

One obvious way to deal with these kinds of problems is to provide the surface owner or lessee not a veto, but some form of compensation for damage occasioned by development of the mineral estate. Where the private surface over federal minerals is involved, Congress has authorized compensation for damage to crops and permanent improvements. Extending compensation to a surface lessee as opposed to a surface owner was authorized by Congress in 1976 for federal grazing permittees when the land in question was devoted to other uses such as mining. But only damages to improvements were covered, and not damages to the ranching operation generally or to the value of the underlying permit. Such compensation schemes can be attacked as unnecessary where the surface rights are subject to competitive bid, because bidders for the surface interest can discount their bids by estimating the likelihood of mining and the potential it creates for interference with the surface interest.

Compensation schemes raise a number of issues: What would compensation cover? How will the amount be determined? And who will pay, the lessee, the federal government, or the states? This last issue might be a critical one. Clawson's proposal (see page 224) calls for a return of 25 percent of the mineral royalty to the surface lessee, not simply as compensation for damage but as a positive incentive to encourage mining. The remainder of the royalty (which, it might be reiterated, currently derives only from the various leasing acts and not from the Mining Law of 1872) under this proposal would be distributed according to current law. Inescapably, current recipients—ordinarily, the federal Treasury, the states, counties, and the reclamation fund (which benefits only the western states)—will suffer some diminution, and are thus likely to resist such an arrangement.

State and local concerns could be allayed by the fact that the states retain the power under the Mineral Leasing Act to tax production of federal minerals, and a recent Supreme Court decision upholding Montana's 30 percent coal severance tax clearly points out one way for states to recoup any reductions in federal royalty sharing arrangements.[28] One reaction to the Supreme Court decision has been, however, a strengthening of sentiment in Congress to limit the size of state severance taxes. Thus the political picture could become even more complicated with the merger of these heretofore distinct considerations—the rethinking of not only the federal lands but also states' powers to tax federal resource extraction.

Finally, where some but not all federal multiple-use lands are intensively managed for particular surface resources, the extent to which mineral exploitation may proceed could affect the location and pace of mineral development on other federal lands not burdened with dominant use leases. This could lead to an expanded set of conflicts, involving not only surface users versus mineral developers but also surface users in one area versus another.

Water resources

Perhaps no single federal policy decision has had as much impact on the western landscape as the decision to sever water from federal land. In a series of statutes enacted shortly after the Civil War, Congress separated management and disposal of federal lands from management and disposal of water, transferring much of its authority over the latter to the states. A few decades later Congress substantially reaffirmed its deference to state water law at the same time it embarked on what became a large federal program to underwrite much of the cost of most large water development projects in the West.

The legal severance of water from federal land, coupled with the federal reclamation program, has had several effects. First, it has allowed the western states to decide when, where, and how most of this water will be used. With federal encouragement and financial assistance, those states have devoted the great bulk of available water to agricultural irrigation. Second, not surprisingly, this water is almost all used on nonfederal lands even though most surface water and some groundwater is, in effect, supplied by precipitation on federal lands, and many groundwater aquifers underlie federal as well as nonfederal lands. In fact, nearly two-thirds of the average annual water yield in the eleven western states comes from federal lands while only about 1 percent of this water is consumptively used on federal lands.[29] Third, in carrying out this grant of authority, each state has developed a substantial administrative machinery to deal with water rights, including the increasingly important function of supervising the transfer of rights from one user or place of use to another.

The exceptions to the otherwise complete federal delegation of authority over water to the states include the following:

1. The federal government has retained, under the so-called *Winters* doctrine, the power to reserve water for uses on reserved federal lands, including federal lands held in trust for Indians.

2. The federal government has retained authority to allocate water among the states, and to control the use of that allocated water within states, especially in connection with federal water projects.

3. Congress has retained the authority to promote navigation and license dams and power-generating facilities on most waterways, including the right in some cases to extinguish private property interests without compensation.

4. Congress has, pursuant to the reclamation program, sought to restrict the use of federally developed reclamation project water for

irrigation to small, individually owned parcels of land (the 160-acre limitation).[30]

Yet despite the furor which continues to be generated about these exceptions, the incontrovertible fact is that even today the water resources of the West are mostly allocated and controlled by state rather than federal law. The federal government has, in other words, only rarely exercised the power it has reserved, or enforced the restrictions it has placed on the exercise of state power, in a way inconsistent with state desires.

Thus, the federal reservation doctrine has only occasionally been invoked to provide significant amounts of water for consumptive use on federal lands. The relatively few Indian reservations which have adjudicated water rights constitute the only prominent exception. While there is a real possibility that future adjudications may allocate large amounts of water to Indians (and, less likely, to non-Indian federal reservations) for consumptive uses, neither the federal government nor the Indian tribes have yet fully embarked on a comprehensive program even to identify, much less to settle, these water rights claims under the *Winters* doctrine.

In fact, the federal government's historic posture has been to subsidize the development of water supplies for non-Indians, giving only scant and, it is increasingly alleged, inadequate attention to Indian and other federal water rights. Furthermore, federal restrictions on the use of reclamation project water like the 160-acre limitation have been largely unenforced and ineffective, either through exceptions created by Congress, by the courts in interpreting ambiguous legislation (in a fashion strongly reminiscent of judicial interpretation of the settlement restrictions in railroad land grants), or by the executive branch as a result of simple lack of enforcement. The recent major liberalization of this limitation by Congress in effect simply ratified the status quo.

The western states being largely arid, there is obviously not enough water for all potential uses. Because agricultural irrigation remains by far the dominant use of water, only a proportionately small amount of land consumes most of the available water. Against these facts, the result of federal policies has been plain: Federal lands generate most of the water supply in the West, but the federal government has not only allowed the western states to use this water on nonfederal lands largely free from federal restrictions, but has, in fact, subsidized such uses through the reclamation program (while largely ignoring the acreage limitation which was originally an important *quid pro quo* for the subsidy). It also should be noted that most western surface streams are fully or overappropriated; that is, claims of the right to use such waters under state law equal or exceed the available flow. Furthermore, groundwater withdrawals in many parts of the West exceed recharge, so that the groundwater is being "mined."

Changes in federal land management would, of course, affect water only to the extent these changes alter existing patterns of water supply and demand. Addressing water supply first, it is difficult to generalize about the effect substantial changes in management of federal multiple-use lands would have on water supply. For one thing, assumptions have to be made about what those changes will be and how they will be carried out. For another, there is a relatively extensive but heavily site-specific, and thus somewhat inconclusive, literature on the connection between watershed vegetation management and downstream effects on surface water supply.[31] Such effects alter not only the amount, but also the quality and the rate of runoff. Much of the literature concerns the effect of grazing and timber harvesting, both of which are primary uses of large areas of federal lands. We know far less about groundwater, of course, but it bears emphasizing in either situation that local conditions and particular management practices seem determinative.

Nevertheless, federal multiple-use lands occupy large areas of most important watersheds in the West, and it takes no great exercise in imagination to envision substantial nervousness on the part of downstream users if the management of those watershed lands is proposed to be altered, especially for the express purpose of intensifying the production of resources other than water. To take just one example, oil and gas leases issued a few months ago on national forest lands in Seattle's city watershed were canceled after vigorous protest on the part of local interests who were concerned mostly about the effect on water supplies.

History suggests this is far from an isolated example, for watershed protection was in fact "the first aim of forest management."[32] At the turn of the century, Congress gave watershed protection primacy equal to timber production in the Organic Act for national forest management. A few years later severe floods served as the catalyst for enacting the Weeks Act, which authorized federal purchase of cutover forestlands in the East. An extreme example was provided by the development and use of hydraulic mining in the Sierra Nevada foothills of California between 1853 and 1884. This caused such destruction downstream that it was eventually outlawed by the courts.

Unless and until downstream users receive satisfactory assurance that the quantity and quality of water supplies will be protected, or that they will be compensated for losses by cash payments or through construction of projects to provide replacement supplies or greater flood protection, we might reasonably expect them to resist significant management changes on federal lands. Such resistance is perhaps even more likely given what might be called the great paradox of water resource management in the West—namely, that although the economic importance attached to water is still relatively low (it being generally conceded that water is substantially underpriced in most parts of the West), emotionally and politically water is extremely important. Since the western states and large users of water in the private sector or quasi-public sector (irrigation districts) historically have been very successful in emphasizing water's political appeal in order to obtain federal funds for maintaining a cheap supply of it, I would expect them to seize upon new proposals for more intensive management of federal lands as another opportunity to extract both further protections and additional federal financial assistance for water resource development. One can only speculate whether the federal government would be willing to pay that price to overcome such resistance.

Further, as noted above, a comparatively small amount of water is now consumptively used on federal lands. Graziers and miners using water on federal lands do so under state law. (Water for timber production is considered separately below.) This means that if Congress continued its pattern of deference to state water law in authorizing changes in management of federal lands, and those changes led to increased demands for consumptive use on federal lands or altered patterns of supply, then state water law would control the adjustment of water rights to these altered conditions. Where a condition of full or overappropriation now exists, of course, new demands or diminished supplies could only be accounted for by transfer from or by other adjustments among existing users. Exactly how this would occur depends upon many circumstances, but I can hazard some speculations. Transfers to meet new demands would be likely to come from irrigated agriculture, which still accounts for about 90 percent of consumptive water use in the West. Mining interests would usually have little difficulty purchasing needed water rights, because water is a comparatively small part of their cost of doing business. New town or recreational developers would probably have greater difficulty, and graziers would have serious problems.

The states would be required to approve such transfers, even where the long-term lessee or new owner of former federal lands in need of additional water found a willing seller. This could place the state in the uncomfortable position of having competing claimants for what water exists, or of having previous users from that source claim injury from its transfer to a new place or type of use. Moreover, in many western states numerous practical and legal barriers exist to transferring water rights from one place or type of use to another. Overall, where water sources are legally overappropriated (not uncommon in the West), any proposed transfer may trigger controversy, delay, litigation, and, ultimately, demands for a general adjudication of all the rights to the source. It is conceivable, in other words, that if proposed changes in federal land use occur that alter patterns of water supply and demand, a substantially disruptive element will be introduced into existing state systems for administering water rights at a time when these systems already are being subjected to substantial stress in many areas of the West.

It is a separate question whether such disruption is, on balance, a negative factor. My own view is that it might actually prove beneficial by promoting greater certainty and accountability in water rights regimes, and by allowing water to move from lower- to higher-value uses. That lower-value uses today comprise a significant percentage of water use in the West seems difficult to dispute. The Colorado River basin, which includes some of the most arid lands in the country, has been described as a "vast feedlot" because most of the water is used for livestock production, even though the region produces a relatively small and easily replaceable portion of the nation's meat supplies.[33]

To the extent that changes in federal land management require alteration in the status quo, the states may not be eager to take on such a potentially divisive task. If the states thwart transfers of water rights for use on federal lands, of course, the objectives of the proposed land management changes could be thwarted.

If Congress wishes to avoid this and to assure users on federal lands sufficient water supplies, it could condemn existing water rights, or give federal land users the power to do so. This is not totally at odds with state law, because some states already give higher-value water users the right to condemn the rights of lower-value users. Precedents for federal condemnation also exist; for example, a few federal reclamation projects have required condemnation of existing water rights. Condemnation might nevertheless be opposed by many western interests who historically have been successful in resisting such federal encroachments, at least those unaccompanied by generous financial settlements.

There is yet another possibility. It is not entirely clear whether the federal government has relinquished its riparian water rights on federal lands which border streams. Thus, graziers, miners, and other transferees of federal lands bordering streams could assert a riparian right to water under federal law (or, possibly, state law in the few western states where the doctrine still obtains). Because the riparian right is not based on priority of actual use, the failure to assert such rights previously may not defeat the claim. The mere mention of asserting a riparian right on federal lands is likely, however, to drive the western states into a frenzy of states' rights rhetoric. The most likely response would be for the federal government to prohibit lessees or purchasers of federal land from asserting riparian rights. This would return the issue to the state systems, raising the questions noted earlier.

Timber interests are in a different position, at least on the national forests which were originally created for timber production as well as watershed protection. There, the federal government has recognized federal water rights in the amount necessary to produce a continuous supply of timber. Dating back to the turn of the century, these federal water rights are usually senior to others in the area. They are also usually upstream from other users, because most national forests are found in the upper reaches of western watersheds. Congress could, of course, require lessees to comply with state water laws, and no doubt the states and existing users would press vigorously for such a result. And if state law were applied, the earlier analysis would apply to timber as well.

To summarize, changes in federal multiple-use land management may have no effect on water resources, especially if such transfers do not substantially alter existing patterns of water supply and demand. To the extent they increase water supplies (such as enhancing runoff) they even could be welcomed. To the extent they significantly increase demand for water or adversely affect water quality, however, they are likely to (1) exacerbate state–federal relations in this delicate area; (2) place the state water-management agencies in the uncomfortable position of determining whether to approve transfers which reallocate water from existing users; (3) increase the likelihood of delay and litigation, including lengthy, expensive general stream adjudications to settle all water claims; and (4) lead to demands on the federal Treasury (or on recipients of federal land) for compensation for interference with existing water rights or for the construction of mitigating water supply projects.

Interstate ramifications are also possible if the water supply is significantly altered by proposed management changes. Interstate allocations of water have been fashioned by compact, Supreme Court decision, or congressional legislation on most important interstate streams in the West. To the extent these may be affected, tensions among states could ensue. Generally speaking, all water consumption on federal lands is counted against a state's allocation of water in such interstate settlements, so that several states might have an interest in major federal lands transactions that could affect interstate flow regimes. We might reasonably expect that federal management changes would not alter water regimes so much as to raise serious concerns on this score, but the possibility cannot be dismissed out of hand.

Still other facets of this complex area deserve mention. Congress might try to meet federal lessees' increased water supply needs—to the extent they would be created—by authorizing them to pump groundwater under federal lands. This would be substantial departure from current policy, but would raise objections only to the extent that these groundwater aquifers also underlie nonfederal lands, are hydrologically connected to surface waters, or are currently being used for other federal purposes (for example, to supply military bases or satisfy Indian water rights claims), so as to affect the supply available to other federal or nonfederal users. This could often be the case, although ascertaining the actual effects could be complicated and expensive. Groundwater hydrology at best is a highly uncertain, variable science, and it is often difficult to determine the dimensions of an aquifer and the movement of water within it. To the extent there is a hydrologic connection between groundwater under federal lands and the use of water off federal lands, a federal decision to allow its lessees to pump significant amounts would exacerbate state–federal tensions. The states would resist such encroachment on their traditional power to allocate and control groundwater, even though the Supreme Court recently has emphasized the federal interest in groundwater regulation. To the extent that groundwater is embraced within the federal water right (likely but not yet definitively settled), it would put the federal timber lessee in a stronger legal position to use it, unless Congress acquiesced under pressures to prevent the assertion of this federal right.

In addition, native Americans hold potentially large, but mostly unquantified, water rights under federal law. These probably embrace groundwater as well as surface water and are superior to most non-Indian rights. Indian reservations in the West are frequently bordered by large tracts of federal multiple-use lands; thus, Indians are likely to be concerned about significant changes in how these lands are managed. In some respects they, like the states, also might resist federal encroachment and the disruption of the status quo. But in other respects they are unique, both because tribes are to some extent independent sovereign entities and because the federal government has a trust responsibility to safeguard their interests. Native American claims may well have to be reckoned with in addressing proposed management changes for federal multiple-use lands, and water may be the place where those claims are concentrated. It is also worth noting that the implications of proposed changes on fishing, hunting, timber harvesting, mining, and grazing rights are of great concern since many Native Americans depend heavily on these activities and could be adversely affected if greater competition resulted from more intensive management of these resources on federal lands.

The mention of fishery interests suggests yet another dimension, because sport and commercial fishermen might have a significant interest in proposed federal land-management changes. To the extent such changes are thought to damage fishery habitat by changing runoff patterns, lowering water quality, raising stream temperatures, and disrupting spawning grounds—effects which, it should be noted, intensive forestry, grazing, and mining have historically been accused of in some areas—there could be resistance from this interest group which commands considerable political power in such areas as the Pacific Northwest.

A final dimension to the water issue involves future projections, if federal lands are made subject to long-term leases for intensive, dominant use management. If these federal lessees perfect substantial new water rights (whether by purchase, transfer, or otherwise) for use on federal lands, will the federal government assert any interest in these rights, will it permit lessees to alienate from the federal lands water rights which have been perfected for use on those lands, or will it allow state law to control the matter? The course chosen could dramatically affect the price the federal government commands for leases and the ease with which lessees could be replaced.

Impacts on state and local governments

Federal land-management policies have always had a marked strand of localism. The political influence exerted by the West on these policies has traditionally been strong and grows with each reapportionment. Important federal land committee positions in Congress are usually held by westerners. Traditionally, most political appointees to key federal executive positions dealing with the federal lands come from the West, where many have served previously at the state level. It is not much of an exaggeration to describe the Department of the Interior as the only regional cabinet department. Career managers of federal lands in the executive branch, reinforced by the decentralized decision-making structure of the agencies themselves, tend to be somewhat more responsive to local and regional concerns than to national ones. Even the Supreme Court, in addressing legal issues relating to federal lands, has been heavily influenced over the years by justices from the West, such as Field and McKenna (California), Sutherland (Utah), Van Devanter (Wyoming), Douglas (Washington), and, currently, Rehnquist and O'Connor (Arizona).

Congress' near-plenary power here deserves reiteration. It could, for example, instantly outlaw, at least prospectively, the application of state fish and game or water laws to activities on federal lands. It could also repeal all mechanisms by which state and local governments receive financial assistance based on federal landholdings, and even prohibit the states from taxing the extraction of resources from federal lands by federal lessees.

These examples are not fortuitously chosen. They are subjects about which state and local governments in the West are extremely sensitive, and embody "rights" that exist basically at the sufferance of Congress, protected only by the political rather than constitutional safeguards of federalism. Perhaps it is precisely because these "rights" lack constitutional protection, and because the political safeguards of federalism are weaker to the extent these concerns are not shared equally by all the states (federal landholdings being smaller and some problems like water supply being very different in nonwestern states), that the West fights so fiercely to perserve them. Witness, for example, the apoplectic character of a typical westerner's reaction to assertion of federal water rights in derogation of state control over, or, as it is mistakenly put, "ownership" of water.

In this same vein, it is useful to remember that the Sagebrush Rebellion gained a measure of political legitimacy because it was the states, and not "greedy" private interests, which carried its banner. (In recognition of this, the Rebellion's opponents have not challenged the states so much as they have tried to characterize their legislatures as dupes for those greedy private interests who, they assert, really want ownership of the federal lands.)

In short, western attitudes, mostly reflected by state and local governments in the region, are important in assessing the feasibility of any proposed major change in federal land-management policies. In fact, history shows that it is increasingly rare for major changes in federal land-management policies to be adopted without the support of, or at least over the objections of, these governments. Thus, the question must always be asked when considering such a proposal: How will the state and local governments view it? And if they are likely to oppose it, what modifications will enlist their support?

In addition to the impact of federal land-management changes on state jurisdiction over water, fish, and wildlife, the most important area of concern will likely be their effect on state and local revenues. Unfortunately, it is difficult to unravel, much less summarize, the complicated and obscure arrangements by which the federal lands generate revenues for treasuries at all levels of government, and under which the federal government provides financial assistance to state and local governments on some federal land resource-based formula. Nonetheless, the matter is highly pertinent to determining the political acceptability of any proposal for fundamental change in the way in which the federal lands are managed.

Some federal land-based revenue-sharing schemes now in effect are tied directly to federal land-derived revenues or resource-extraction rates. For example, almost all western states directly or indirectly now receive 90 percent of the royalties collected from federal mineral leases within their borders. There are similar arrangements for dividing the proceeds from timber sales and grazing fees. If federal revenues from these activities increase, so automatically do state and local revenues. If more intensive management of federal lands increases resource outputs, it will have positive effects on state and local revenues, and thus would likely be supported by these governments.

Other considerations complicate the picture, however. Some financial arrangements are based merely on the presence of federal land regardless of how it is managed. One example—which, although it has resulted in many millions of dollars of additional federal payments to states with large federal landholdings, is curiously absent from most discussions of federal land-based revenue sharing—is found in the provisions of the Federal Highway Act which increase the federal contributing share for construction of interstate and other federal highways in a state based solely on the percentage of public and Indian landholdings in that state.[34] Congress added to this and other special programs in 1976 by enacting the Payments in Lieu of Taxes Act. This, as its name reflects, seeks to compensate the states for the tax immunity of federal lands by authorizing direct federal payments to local governments that are based on the amount of federal landholdings within their jurisdictions.

To the extent that alterations in management practices do not transfer title out of federal ownership, it might be argued that such revenue-sharing schemes—most of which officially depend only upon federal title—should not be affected. While literally correct, this view is naive, for these kinds of arrangements are already coming under increasing attack as being overgenerous, particularly in this era of large federal deficits. The Advisory Commission on Intergovernmental Relations (ACIR) and the General Accounting Office (GAO) have recently completed studies reaching essentially identical conclusions: such federal programs are inequitable, inefficient, and, most significantly, overcompensate most western state and local governments.[35] The GAO, for example, concluded that various federal land-based payments in fiscal year 1977 for six western states exceeded actual tax equivalency (that is, the revenues generated if states had simply taxed federal lands at prevailing rates) by nearly $200 million (and GAO ignored federal highway funds). The ACIR found that the presence of federal lands added little if any cost burden to state and local governments, thus undercutting the rationale behind most of these revenue-sharing and compensation schemes.

Another ingredient in this fiscal mix is that states today are substantially unfettered in their ability to tax resources extracted from federal lands. State coal severance taxes recently gained considerable notoriety when the Supreme Court upheld Montana's 30 percent severance tax, which applied almost exclusively to federal coal. The Court relied in part on the fact that Congress could always prevent burdensome state taxation, and in response legislation has been introduced by representatives from coal-consuming regions to limit state taxing power. As noted earlier, proposals to alter management of multiple-use lands in any significant way could fuel that emerging discontent.

Other financial impacts

In addition to intergovernmental fiscal arrangements, private users of federal lands sometimes derive special financial benefits from federal land policies. Not only is it widely believed that federal grazing fees are below market value, for example, but Congress has required that half those fees be plowed back into federal rangelands in the form of range "improvements." And these investments are in addition to the several hundred million dollar, twenty-year program of federal range restoration authorized by Congress in the Public Rangeland Improvement Act of 1978.[36] Almost all federal oil and gas leases are issued noncompetitively, for a small filing fee, even when the lands being leased are known to be excellent prospects for hydrocarbons. Hardrock miners pay no royalty or rental and can purchase federal land for $5.00 or less an acre. The Forest Service sells timber in some regions for prices which, it is alleged, are below the cost of producing the trees. Recreationists, including millions of hunters and fishermen, are usually not charged a federal fee for use of federal multiple-use lands.

Such arrangements would be scrutinized as part of any general re-evaluation of the federal lands. Would Congress continue to perceive a need for federally funded range improvements if graziers were given a long-term, legally protected property interest in federal lands? The answer is quite possibly no, particularly if a principal motive for the change is to provide incentives for private, rather than public, investment in the rangelands. Therefore, we might expect private users as well as state and local governments, acting in accordance with sound economic principles, to maximize net gains for themselves by resisting change. It is worth recalling that the PLLRC's proposal to sell prime federal rangeland to grazing permittees received a tepid response from livestock interests, probably because they realized they could not obtain use of the lands as cheaply by outright purchase as they do now under permit.[37]

For these reasons, state and local governments and private users of federal lands may well find themselves on the defensive should Congress ever seriously address the financial implications of federal multiple-use land management, as it almost certainly would if it sets its sights on increasing management efficiency and resource outputs. Indeed, there is no assurance that these existing financial arrangements would survive unscathed, especially given current federal fiscal straits. Nevertheless, it is worth recalling that the arrangements now in place have not been plucked from thin air, but instead have evolved through gradual accommodation and adjustment in Congress. Even as federal deficits were mounting, the trend of federal legislation since the turn of the century clearly has been toward more generous financial arrangements with nonfederal entities.

In recent years, there have been numerous and bipartisan attempts by the Office of Management and Budget and others in the federal establishment (supported by many PLLRC recommendations) to wring more revenues from, or to reduce subsidies to users of, the federal lands, and to reduce federal lands-based financial assistance to state and local governments. Battles over such issues as fees for admission to federal recreation areas, competition in and levels of timber sales on national forests, appropriation levels for payments in lieu of taxes, grazing fees, mineral-leasing competition and royalty rates, and federal license fees for hunters and fishermen have been periodically waged and largely won by state and local governments and public land users. The federal pot is not bottomless, of course, and a few attempts to enlarge nonfederal shares in federal land revenues have been thwarted, at least temporarily. Thus an effort by coastal states to tap a share of the lucrative federal receipts from Outer Continental Shelf oil and gas leasing has so far failed in Congress, though support for it is growing. (Interestingly, the state and local governments across the country already indirectly share in those receipts, a substantial share of which is recycled to them through the Land and Water Conservation Fund to underwrite the purchase of property for recreational purposes.)

The increasing frequency with which these fiscal battles are being waged underscores the opinion offered here that the fiscal implications

of proposed changes will be crucial in determining the political acceptability of proposed changes in federal lands policies.

Taking a broader view, a thoroughgoing cynic might suggest that the
objective of those in the West who embraced the Sagebrush Rebellion
was not to wrest ownership of the lands from the federal government,
but instead was another attempt on the part of the West to use the large
federal landholdings as a lever for extracting financial concessions from
Washington. These lands have, the cynic would say, proved to be an
effective crowbar, wielded over the years by western politicians of all
leanings, to pry open the doors of the federal Treasury.

In the early 1930s, for example, the West rejected President Hoover's
offer of large tracts of federal lands, not only because the most lucrative
parts (the minerals) were not offered, but also, the cynic would maintain,
because the states recognized that the federal status of these lands furnished a basis for numerous special programs of federal financial assistance that constituted an umbilical cord for the western economy. A
substantial restructuring of federal lands (especially amid a general re-
trenchment in federal expenditures) would sever that cord and eliminate
any useful rationale for special treatment. Thus Rep. Edward Taylor of
Colorado, previously an advocate of federal land grants to western stockmen, eventually sponsored the legislation bearing his name which ce-
mented federal retention of those rangelands. More recently, George
Will has suggested that a welfare mother is the soul of self-reliance
compared to a westerner who receives federally subsidized range privileges, water, and various other benefits tied in one way or another to
the existence of federal landholdings. And when the Attorney General
of Nevada came calling on the Department of the Interior shortly after
the Nevada legislature had officially claimed ownership of the federal
public lands within its borders, he was concerned not with developing
an orderly method for testing the constitutionality of the claim, but
rather with obtaining assurance that federal payments based on federal
title would continue to flow uninterrupted by Nevada's claim that it,
and not the federal government, owned these lands.

Developing the cynical view a bit further, once westerners raised the
ownership issue largely for purposes of extracting financial concessions,
it was inevitable that those advocating "privatization" of the federal
lands on pure policy grounds would seize the issue for their own. But,
the cynic also might suggest, most westerners are not particularly interested in economic efficiency, or in using cost–benefit analyses and
other marketplace tools to determine the wisdom of governmental investment in federal lands, or in using open competition to determine
the price of those lands or their resources. (The National Water Com-

mission recommended, and President Carter attempted, reforms along those lines in federal water policy, and neither won any popularity contests in the West or in Congress.[38]) Thus the Rebel position now embraces in uneasy alliance, as political movements often do, interests with common complaints but seriously diverging objectives.

Whether such an alliance will hold together long enough to realize some success in Congress is impossible to predict. The cynic would say, however, that if a major restructuring is ever to emerge from the political process, financial concessions will have to be made to state and local governments and federal land users. Ultimately, the issue will become whether the game is worth the candle—whether these financial concessions will prove so large as to destroy the efficiency-related goals of reform. Unless the conflicts between efficiency and maintenance of federal largess can be resolved through adroitly fashioned compromise, in other words, the movement for change and, more important, the reasons for change, will founder.

Conclusion

The assumption underlying most of the observations offered here is that federal multiple-use land management is perceived as a zero-sum game by the complex array of interests directly concerned with it. These interests include not just the most obvious users of the land, such as ranchers, loggers, miners, and recreationists, but also numerous others affected in some relatively immediate way by how the lands are managed, such as state and local governments, those dependent upon water furnished by these lands, and Indians on nearby reservations. The prevailing perception is, in other words, that while some interests will benefit, others will inevitably suffer from any significant change in that management.

Throughout this volume arguments have been forcefully presented that some changes can benefit most, and perhaps all, of these interests. Abandoning the nondeclining evenflow policy of timber output from national forest lands, for example, is presented as a way to provide both more timber and more wilderness. The same rationale underlies the arguments for privatization or long-term leasing. Nevertheless, the actual effect of proposals promising general benefits for all is not sufficiently certain to overcome the widely shared belief that some interests must inevitably lose. So far, in other words, advocates of even the most powerful cases for specific reforms have not carried their heavy political

272

burden of persuading key interests they will benefit from an alteration in the status quo.

The key interests concerned with federal multiple-use land management are relatively numerous these days and, perhaps more important, each interest commands enough political power to exercise veto power against proposals for major change which it finds unacceptable. As a result, existing arrangements for use of these lands have grown up largely incrementally, by piecemeal, transitory compromise. Though my purpose here has not been to serve as an apologist for the status quo, in a sense that is the burden of my presentation. By leaving federal land management to the vagaries of the political process, we have erected a patchwork structure which is, like the tax code and the political process they both reflect, full of inertia and Byzantine interrelationships.

The result is likely to be a stalemate on bold proposals for reform, and continuing incremental change. It usually has been this way, although the dramatically increased political power wielded by recreationists and preservationists and the enlarged role of the courts in recent years have further diffused the power to influence federal land management policies, and have enlarged the dimensions of the balancing of interests which shapes those policies. And, I believe, it will continue to be so until the right combination of ideas, people, and triggering events can be found to dramatically shift political power among those diffused interests. Perhaps "privatization," Ronald Reagan, and the Sagebrush Rebellion are such a combination, but the signs are not, I think, particularly favorable to their cause.

Notes

1. The Surface Mining Reclamation and Control Act of 1977, 30 U.S.C. (United States Code), sect. 1201–1328.
2. The Weeks Act of 1911 (eastern national forests), 16 U.S.C., sect. 513–519.
3. The Bankhead–Jones Farm Tenant Act of 1937 (national grasslands), 7 U.S.C., sect. 1010–1012.
4. Statehood land grant restrictions are discussed, with references, in *Lassen* v. *Arizona*, 385 U.S. 458 (1967). See also Note, "Administration of Grazing Leases of State Lands in New Mexico: A Breach of Trust," *Natural Resources Journal* vol. 15, p. 581 (1975).
5. The Recreation and Public Purposes Act, 43 U.S.C., 869.
6. The Hetch-Hetchy grant and litigation is recounted in *United States* v. *City and County of San Francisco*, 310 U.S. 16 (1940); *City of Palo Alto* v. *City and County of San Francisco*, 548 F.2d 450 (9th Cir. 1977); and *City and County of San Francisco* v. *United Airlines*, 616 F.2d 1063 (9th Cir. 1979).

7. The legal history of the Railroad Land Grants is exhaustively discussed in "Railroad Land Grants—Legal Analysis," prepared by the Office of the Solicitor, Department of the Interior (Oct. 14, 1976).

8. Paul W. Gates, *History of Public Land Law Development* (Washington, D.C., Zenger, 1968).

9. The Stock-Raising Homestead Act of 1916 was found at 43 U.S.C., sect. 291–302 (1964). It was repealed by the Federal Land Policy and Management Act of 1976.

10. U.S. Public Land Law Review Commission, *One Third of the Nation's Land* (Washington, D.C., Government Printing Office, 1970).

11. C. Brimmer, "The Rancher's Subservient Surface Estate," *Land and Water Law Review* vol. v, no. 1, 1970, pp. 47–64.

12. The surface owner consent provision of the Surface Mining Reclamation and Control Act is 43 U.S.C., sect. 1304.

13. The Small Tract Act of 1938, as amended, was found at 43 U.S.C., sect. 682a (1964). It was repealed by the Federal Land Policy and Management Act of 1976. The secretary's refusal to promulgate regulations or allow mineral activity in such areas was upheld in *Dredge Corp.* v. *Penny*, 362 F.2d 889 (9th Cir. 1966).

14. The Federal Land Policy and Management Act of 1976, 43 U.S.C., sect. 1701–1782.

15. The Mineral Leasing Act of 1920, 30 U.S.C., sect. 181–262, as amended in part by the Federal Coal Leasing Act Amendments of 1976, 30 U.S.C., sect. 201–209.

16. The Taylor Grazing Act of 1934, 43 U.S.C., sect. 315–315r.

17. Offroad vehicles were the subject of President Nixon's Executive Order no. 11644 (1972). Implementation was criticized in Sheridan, *Off-Road Vehicles on Public Land* (Council on Environmental Quality, 1979).

18. Ski area permits on national forest land are governed by 16 U.S.C., sect. 497.

19. The Alaska pipeline litigation is *Wilderness Society* v. *Morton*, 479 F.2d 842 (D.C. Cir.), *cert. den.* 411 U.S. 917 (1973).

20. The Wilderness Act of 1964, 16 U.S.C., sect. 1131–1136.

21. The Sustained Yield Forest Management Act of 1944, 16 U.S.C., sect. 583–583i.

22. Glen O. Robinson, *The Forest Service* (Baltimore, Md., Johns Hopkins University Press for Resources for the Future, 1975).

23. Paul J. Culhane, *Public Lands Politics* (Baltimore, Md., Johns Hopkins University Press for Resources for the Future, 1972).

24. William C. Patric, *Trust Land Administration in the Western States* (Denver, Public Lands Institute, 1981).

25. Robert W. Swenson, "Legal Aspects of Mineral Resources Exploration," in Gates, *History*.

26. The Mining Law of 1872, 30 U.S.C., sect. 21–54.

27. The Acquired Lands Mineral Leasing Act, 30 U.S.C., sect. 351–359.

28. The Supreme Court decision on Montana's coal severance tax is *Commonwealth Edison* v. *Montana*, 453 U.S. 609 (1981).

29. PLLRC, *One Third*.

30. On water rights and federal lands, see Meyers and Tarlock, *Water Resources Management* 201–237 (2d ed., Foundation Press, 1980).

31. William E. Sopper, *Watershed Management*, prepared for the National Water Commission (Springfield, Va., National Technical Information Service, 1971).

32. Robinson, *The Forest Service*.

33. Philip Fradkin, *A River No More* (New York, Alfred Knopf, 1981).

274 PART IV: INTERMEDIATE POSITIONS AND SPECIAL PROBLEMS

34. The Federal Highway Act's revenue sharing provisions are found at 23 U.S.C., sect. 120.

35. U.S. Advisory Commission on Intergovernmental Relations, *The Adequacy of Federal Compensation to Local Governments for Tax Exempt Federal Lands* (Washington, D.C. Government Printing Office, 1978); and U.S. General Accounting Office, *Alternatives for Achieving Greater Equities in Federal Land Payment Programs* (Washington, D.C., Government Printing Office, 1979).

36. The Public Rangeland Improvement Act of 1978, 43 U.S.C., sect. 1901–1908.

37. Robinson, *The Forest Service*.

38. National Water Commission, *Water Policies for the Future* (Washington, D.C., Government Printing Office, 1973).

12

Ideology and Public Land Policy
The Current Crisis
ROBERT H. NELSON

Historically, public policies have closely followed broader trends of ideology. The nineteenth-century policy to dispose of the public lands fit the prevailing classic liberal ideology and the general expectation of a small role for government, especially the federal government. The turn to federal retention of the public lands occurred as a key element of the Progressive movement; indeed, conservationism represented the application of progressive ideology to the field of natural resources. The ideas of multiple-use management formulated in the past several decades are an application to the public lands of the dominant ideology of these decades, labeled by Theodore Lowi as "interest group liberalism."[1] In this concept the governing process is seen as the achievement of an acceptable balance among contending interest groups on the public lands—the multiple users of the lands. The current interest in rethinking federal land ownership is part of a broader assessment of the roles of government with many of the same roots as the deregulation movement.

Federal land ownership is an anomaly in the American system. It violates the principle to decentralize responsibility to state and local government, as well as the even more central principle to rely on the private sector for most matters. Many arguments for public land ownership would apply equally well to existing private property—for example, the need to conserve resources for the future. If such arguments were actually applied uniformly throughout the American economy, the result would be a virtual program for nationalizing American industry.

A valid defense of existing public land ownership thus must show how public lands and resources are distinguished from lands and resources for which existing private ownership is also considered appropriate. In particular, a concept that justifies wide public ownership of land in the West should also be able to explain why similar land in the East instead should be privately owned.

There has been surprisingly little examination of such fundamental questions. The inclination has been to leave the current division of public and private land ownership unchallenged, moving on to more "pragmatic" subjects. There has always been an element of the American character which disdains matters of social theory and ideology. Yet, it is impossible to function without a guiding sense of direction provided by ideology. To neglect such matters is not to banish them but to make them implicit rather than explicit. It also runs the risk that attention will constantly be diverted to symptoms, when a basic cure could be achieved. Rather than always bailing the boat, a new design may be preferable.

Marion Clawson's chapter 10 is to be commended—as is this workshop—for its willingness to reexamine the question of public land ownership and to raise fundamental alternatives. These matters have not really been reassessed in any comprehensive way since the Progressive era, although many major changes have occurred since that time. My purpose in this chapter is to indicate how basic economic and political ideology is intertwined with questions of possible disposal of public land to private ownership, transfer of public land to the states, and other alternative tenure arrangements.

The nineteenth-century preference for the private market involved not only a confidence in the workings of markets but a strong aversion to government. Government was feared as a threat to political liberty— a reaction in part to the lingering memory of British rule. The capabilities of government were also questioned; American democracy was often a rough-and-ready affair, hardly fit for administering details of much complexity or sophistication. The overcoming of these attitudes and the laying of the foundations for a much larger governmental role fell to the Progressives. As noted above, current public land institutions were largely a product of the Progressive movement at the turn of this century.[2]

The Progressive Thesis and the Interest-Group Liberal Antithesis

Progressives sought to introduce to government the same methods of modern management that were then transforming the private sector from a world of small entrepreneurs to one of large corporations. They generally advocated a much more businesslike and efficient basis for government administration. In order to protect professional administrators from political interference, Progressives from Woodrow Wilson on asserted strongly that there should be a strict separation of government administration from politics. The political process would set the broad policy objectives; after that, however, professional administrators applying sophisticated management skills would take over and be responsible for implementing these objectives.

The much greater scope for government advocated by Progressives did not reflect any newfound confidence in democratic decision making. Progressives, in fact, remained profoundly distrustful of democratic politics. However, they believed that government could be isolated from its influence and left to the experts.

The idea that government administration could be truly scientific played a critical role in Progressive thinking. Fearful as always of abuses of government power, Progressives offered protections by means of the assurance that administration would be carried out in a scientifically objective manner. If necessary, the judiciary would enforce this requirement. Progressive institutions such as the independent regulatory commission were designed to insulate administrators from political interference. In this concept government administration should involve little, if any, leeway for the individual feelings, values, or intuitions of the administrators, because science and technology would reveal the one valid answer. Progressives regarded special-interest groups with particular disfavor; such groups sought to distort management decisions to their own advantage and against the valid goal of "the public interest."

Retention of public lands was desirable, according to Progressive and conservationist ideology, for two primary reasons. First, some conservationists, such as Gifford Pinchot, contended that the government would do much better than the private sector in achieving the scientific management of natural resources. In part, Pinchot was reacting to the recent rapid depletion of wildlife, forests, and other natural resources by private individuals and companies. The private sector of the time was considered too short-sighted and too parochial in outlook to achieve a rationally efficient development of natural resources.

But, more critically, it seemed, at least from the evidence of the private sector, that scientific management went hand in hand with large size. As the trust-busting episodes showed, another element of Progressivism was its antagonism to large concentrations of private power, another possible threat to American democracy. If scientific management, in fact, required large organization, then the public sector was preferable to the private sector—or so it seemed to Progressives. Rather than create one or more new Weyerhaeuser empires, Progressives preferred to create the Forest Service.

On the public rangelands the first national Public Lands Commission in 1880 had noted that efficient ranching use of the range required landownerships of 2,500 acres or more—a conclusion that followed the analyses of John Wesley Powell. However, Congress could never swallow disposal to ranchers in such large amounts. It persisted for fifty more years in pursuing the vision of the small homestead, despite overwhelming evidence that 160 acres or even 640 acres was far too small. The Taylor Grazing Act in 1934 finally ended homesteading. The Congress in effect determined that, if small private ranches were not economic for rangeland use, the lands were better retained in public ownership with grazing controlled on a permit basis.

The Progressives thus hoped both to achieve the efficiency of scientific management for government lands and at the same time to prevent large landholdings that might frustrate traditional American egalitarian and democratic ideals. As Samuel Hays has noted with respect to Theodore Roosevelt's thought, these dual objectives were in many respects contradictory: "Roosevelt's emphasis on applied science and his conception of the good society as the classless agrarian society were contradictory trends of thought. The one, a faith which looked to the future, accepted wholeheartedly the basic elements of the new technology. The other, essentially backward looking, longed for the simple agrarian arcadia which, if it ever existed, could never by revived."[3] Moreover, this element was not the only feature of Progressive thought that was seriously flawed. In particular, the critical assumption that the administration of government could be kept strictly separate from politics was undermined regularly by events.

Indeed, in the post-World War II period many close observers of government found an altogether different process. Leading economists and political scientists such as Herbert Simon, David Truman, and Dwight Waldo established their reputations as critics of progressive ideology.[4] Rather than pursuing any identifiable "public interest," government behavior was actually determined largely by competition among special interests. Many post-World War II social theorists considered this governing process not only a valid description of the real world, but as a healthy development; Galbraith, for example, lauded the happy balance achieved in a system of "countervailing powers."[5] Rather than the classic liberal competition of the free market, proponents of this "interest group liberalism" instead applauded competition in the political arena as the basic mechanism for resource allocation.

As noted above, on the public lands this philosophy is reflected in the tenets of "multiple-use" management which seeks a balance among all user groups. The Multiple Use and Sustained Yield Act of 1960 created a formal statutory mandate for multiple-use management of the national forests. Recent statutory requirements for public participation in the decision making for the public lands seek to guarantee that each user group receives a fair chance to influence land management. Leshy's chapter illustrates the current prevailing view that public land policies really have been, and very probably will continue to be, made by interactions among competing interest groups. Implicitly at least, Leshy seems to suggest that there is really no other way (see chapter 11).

The ideologies of Progressivism and interest-group liberalism have in common that they specify, and to some degree justify, the role of government. Beyond this, they are mostly antithetical. Progressivism seeks to banish interest groups; interest-group liberalism puts them at the center of the governing process. Progressivism seeks objective government decisions reached by technical experts using scientific methods; interest group liberalism simply favors an equilibrium of interest-group pressures. Progressivism advocates extensive planning; interest-group liberalism specifies a brokering among interest groups as the primary mechanism for decision making.

To be sure, elements of both ideologies are often intermingled in government policy pronouncements—a source of much confusion. Indeed, the original Progressive ideals have never been formally abandoned in public land management. For the record Congress still speaks frequently of scientific management of public lands in the public interest. Recent legislation such as the Forest and Rangeland Renewable Resources Planning Act (1974), the National Forest Management Act (1976), and the Federal Land Policy and Management Act (1976) blends elements of old-style Progressivism (for example, formal comprehensive planning) and of interest-group liberalism (for example, public involvement and participation). However, close observers of the implementation of such legislation often find Progressive elements to be mainly window dressing. Clawson, for example, notes the disappointing results of the expanded formal planning on the national forests following enactment of the RPA and NFMA. Instead, the key decisions are actually reached in response to interest-group pressures.

The Challenge to Interest-Group Liberalism

Having displaced progressive ideology, the ideology of interest-group liberalism itself is now increasingly challenged. A major concern is that interest-group compromise seldom produces efficient use of resources. Effective planning is similarly impractical in such a governing process. Clawson has been a long-time critic of the inefficiency of public land management, once describing the management of the national forests as "disastrous" for its failure to achieve the full economic potential.[6] For Clawson, as for most economists, a basic problem is that political decision making lacks clear objectives and provides no strong incentives for efficient use of public land or other resources.

Another group of critics finds that interest-group bargaining too often becomes a "zero-sum game," where efforts are devoted primarily to maximizing individual benefits ("rents") from government actions. An analogy can be drawn between the federal Treasury and a common pool resource. The individual claimant on the Treasury is very unlikely to be large enough to affect tax rates and thus his own tax payments. His incentive thus is to maximize his own claims, without much regard to costs. The overall result, however, of all people behaving this way is to "overallocate" the federal Treasury, causing budget deficits and inflation. Budgetary overallocation is the public finance version of overgrazing the commons.

Government decision making by interest-group competition tends to favor concentrated over diffuse interests. As a result, government agencies often become the captive of interest groups most affected by their actions. The big losers include the diffuse body of national taxpayers. The evidence of widespread "regulatory capture" played a major role in promoting the movement to deregulate the Interstate Commerce Commission, Civil Aeronautics Board, and other government regulatory agencies. The recent proposal for a balanced budget amendment to the Constitution—whatever its specific merits—reflected a wide perception that existing political institutions are incapable of controlling interest-group pressures.

Such problems are all found in public land management. There is a low premium placed on efficiency, resulting in high management costs. At the same time interest-group pressures steadily expand the services of the public lands, but prevent significant charges from being imposed for these services. The consequences are that public land management incurs large losses that must be paid by the general taxpayer—much as the overall budget tends to be in deficit. The management costs of the Forest Service, for example, were $2.1 billion in 1981, greatly exceeding the $1.3 billion in revenues, despite the availability to the Forest Service of vast mineral, timber, and other natural resources of high value. The BLM grazing program yielded only $25 million in grazing fees in 1981, compared with costs (fully accounted) of more than $100 million.

Federal minerals management involves few costs (since the minerals are leased) and brings in large revenues. Yet, even in this case, the federal government itself achieves little benefit. Fifty percent of mineral leasing revenues go directly to western states, and another 40 percent are paid into the Reclamation Fund for support of western water projects (although such payments probably substitute for other federal funds.) When federal management costs and corporate tax consequences are taken into account, the federal government in many cases may even lose money on its mineral leasing activities. Most recently, the federal government effectively conceded a portion of its coal revenues to rancher owners of private surface above federal coal. Disposition of hardrock minerals under the mining law does not generate any federal revenues.

Examination of public land investments and production levels shows numerous economic inefficiencies. The inability of many western water projects to pass elementary economic tests has long been noted. More recently, similar criticisms have been directed at Forest Service timber harvesting: timber investments are based on invalid economic calculations that overstate benefits because of the "allowable cut effect"; timber harvesting occurs over wide areas where harvesting costs exceed revenues generated; and timber harvests are kept too low because of "even flow" and other constraints that have little, if any, economic justification. In chapter 4 Hagenstein estimates that timber prices in the Pacific Northwest would be reduced much below forecast levels if timber-harvesting levels on the national forests were determined in response to market incentives. As Hagenstein notes, Forest Service policies thereby advanced greatly the interest of private timber owners, a result that accords well with interest-group liberal concepts. Perversely, many of the inefficiencies of timber management promote harvesting in environmentally sensitive areas, compounding the social losses incurred.

Charges imposed for the use of public lands follow the predictable pattern of government decision making in response to interest-group pressures. Users are politically concentrated and thus defeat efforts to impose user charges that would relieve some of the costs now imposed on highly diffuse national taxpayers. Counting the many National Park System units that have no charge at all, the average fee in 1981 for park visits was a nickel. Hunting, fishing, and other recreational use of BLM and Forest Service lands does not involve any public charge. However, many private landowners whose lands control access to public lands collect substantial hunting fees (up to $50 per day or more) for granting of such access. Ranchers are willing to pay large "permit values" for the privilege of buying public forage at the existing grazing fee levels.

The results of interest group pressures tend to allocate *de facto* rights to use of public lands among various user groups—in his chapter Hagenstein refers to these rights as "restrictive arrangements." Historic assignments of uses tend to harden into permanent commitments. Ranchers thus are assured that certain levels of grazing on adjacent public lands will continue; wilderness users are similarly assured that certain areas will be kept free of competing uses. In the case of water the assignment of rights based on historic use has been made *de jure* as well as *de facto*. In one perspective the fundamental problem of interest-group liberalism is that the creation of property rights is usually not complete; the rights cannot be transferred by selling them to someone else. Hence, efficiency-promoting transfers of rights to higher-value uses do not occur. It has recently been suggested, for example, that wilderness proponents should be able to sell oil- and gas-drilling rights in wilderness areas.[7] The idea is that the revenues from such sale could be used to further wilderness purposes that would be more valuable than any damages to wilderness caused by oil and gas drilling. Other examples in the water, grazing, and timber sectors could easily be given, showing inefficiencies resulting from limits on the ability to reassign existing user entitlements.

There is widespread agreement about many of the specific criticisms of public land management—the high cost, the user charges well below fair market value, the inefficient investments, the lack of economic sophistication, the much greater responsiveness to local and other concentrated interest-group pressures. The critical difference of opinion concerns the solution to the ailment. Are such problems the result of major failings endemic to the current system of public land management, or are they simply the normal loose ends in any large and complex enterprise? Should we be concerned that the jar is half empty or thankful that it is half full? The answers to such questions ultimately depend on the alternatives. Clawson ably presents the basic alternatives to the current public land system. In the remainder of my comments I offer some reflections on each of the alternatives.

Improved Current Management

Proponents of improving current management of the public lands fall into two camps. If not in quite so many words, one group asserts that current problems actually reflect a failure to implement the progressive and conservationist prescriptions. Public land management needs to be insulated more fully from political interference, to apply the latest scientific and technical methods, and to be turned over to the professional experts.

The leading proponents of this reform strategy currently are economists.[8] In their view, the original conservationists failed to achieve their aims partly because they failed to understand that public land management is fundamentally an economic problem—including all nonnmarket as well as market outputs in the economic calculations. Lacking an economic foundation, conservationism became not a true science but an emotional crusade—advocating a "gospel" in Sam Hays's term. But if the scientific methods of economics can be introduced, it will now actually be possible to turn land management over to experts (that is, economists). In essence, often using technical economic methods, the true social benefits and costs of possible public land actions can be calculated, and the one alternative selected which produces the highest net benefits. This is the approach described by Clawson in his discussion of possible improvements to current land management.

Frequently without realizing that it is an opposite prescription, another group proposes to improve current land management by making it more faithful to the precepts of interest-group liberalism. The objective is not to exclude politics, but to make land management even more political. Such diagnoses of current management failings emphasize two elements: (1) an attempt to exclude legitimate interest groups (such as environmentalists); and (2) heavy-handed and clumsy tactics in seeking to broker politically viable compromises among the contending interests. Critics of the Forest Service thus assert that it remains rigidly wedded to archaic concepts of independent management by professional experts and seeks to exclude valid public involvement. Other critics of federal coal leasing in the Interior Department assert that it pays insufficient attention to the concerns of western states, environmentalists, and other concerned interest groups. Chapter 5 by Michael Harvey reflects a common view regretting the intransigency and divisive behavior of interest groups on both sides of many public land issues. Harvey suggests that current problems could be solved if only contending interest groups would be more reasonable, speak more softly, and bargain together in better faith.

Proponents of both these views are curiously lacking in historic perspective. Neither refers to the debates that have gone on for many years with respect to the ideological merits and failings of Progressivism and interest-group liberalism. Indeed, the fundamental disageements between these two ideologies—with respect to matters such as the roles of expert professionals, use of scientific methods, and place of politics—are scarcely noted. Lacking this historic context, there is no good explanation of why public land management has worked poorly in the past or why matters will be different in the future. There is a strong flavor of wishful thinking (or, more cynically, it may be a good way of avoiding change and preserving a status quo that is satisfactory to many).

The actual prospects for reviving the progressive conservationist approach appear dim for several reasons. First, the prospects for greater separation of land management from politics are no better than they have been. Indeed, they are probably poorer; public willingness to defer to experts depends on confidence in the scientific methods and objectivity of their conclusions. But in recent years there has been a notable erosion of public confidence in professional expertise. In many respects current public attitudes are probably well founded; in my own view the social sciences in past years have made many excessive claims and have been slow to grasp their real limitations.[9]

The prospects for more harmonious bargaining and more productive compromises among interest groups also seem poor to me. The incentives are simply wrong; the squeaky wheel still gets the grease. Much as bad money drives out good, loudly aggressive interest groups drive out soft-spoken moderate advocates. The internal politics of interest groups of all political complexions also seem to reward militants and to penalize moderation.

In most cases implicitly, proponents of interest-group liberalism have assumed that interest groups would show enough common sense that one way or another they would be able to agree on measures that enlarge the social pie for everyone. The process of debate over the distribution of the pie would not prevent agreement on measures that would make everyone better off. In recent years, however, this image has seemed less valid; as Leshy comments, interest-group bargaining has instead seemed more a zero-sum game where each interest group takes for granted a pie of fixed size. Indeed, such behavior by interest groups individually is "rational," because any one interest group by itself is unlikely to have an effect on the amount of total social product. The cumulative result, however, may well have been significant social loss, represented in part by inefficient government actions.

As indicated above, the concept of governing by seeking the equilibrium point among the vectors of all interest-group pressures is troubling in many ways. It offers no standard for judging whether a government action is any better or worse than any other—other than the fact that political agreement has been reached. Reminiscent of Voltaire's Pangloss, any course of government is rationalized because it simply shows the current distribution of political power. What happened to such old-fashioned ideals as equity, liberty, justice, and efficiency in government?

The best case for focusing on incremental improvements in current public land management rests on a view that Lindblom described in a famous article, "The Science of Muddling Through."[10] If not exactly a science, muddling through is unquestionably a government tradition of long standing in the United States. Moreover, bitter experience has taught us to be very skeptical of any proposals for radical change; the consequences are seldom what the architects had in mind.

Carried to its full logic, however, such incrementalism becomes a formula for stagnation. The hope for broad change of any kind seems to involve a symbiotic relationship between idealism and incremental thinking. The radical visions of idealists are very seldom realized, but create the social tensions and provide the ideas which bring incremental movement in that direction. Currently, the most radical change being proposed for the public lands is the outright divestiture of most of them to the private sector.

Sale of the Public Lands

One of the major developments in economics over the past several decades has been the invasion by economists of the territory formerly reserved to political scientists—although once all such matters were the domain of "political economy." The students of government coming from the discipline of political science generally observed a tradition that emphasized descriptive (as opposed to prescriptive) analysis of political institutions. Economists brought two distinctive elements, a long tradition of offering strong policy recommendations (mainly for small government and reliance on the market) and an assumption that political actors behave in a self-interested way (an analog to the standard assumptions about private market behavior). Applying traditional economic reasoning in the context of political institutions, the conclusions have not been favorable to these institutions. Bureaucratic incentives are to expand bureaucracy, not to provide an optimal level of services most efficiently. Most seriously, if government is employed as an instrument for maximum self- (or group) enrichment, broader social objectives may be lost. Large resources may even be devoted to the competition for government largess. It seems that effective government depends on a large dose of altruistic and genuinely public spirited behavior—by both government administrators (bureaucrats) and the citizens affected by government actions. Yet, the supply of such behavior varies greatly in different societies and circumstances; indeed, the trend in the United States in recent years has perhaps been downward, as people actually behave more like the economic man of economic theory. Proponents of new government responsibilities often fail to address such issues.

Another recent development in economics has been a greater attention to the problems of information generation. Rather than a world of perfect information and equilibrium, the real world may be characterized by widespread ignorance and rapid change. In such a world the rewards to planning and conscious design—scientific management—may be less than had been thought. Chance and happenstance instead may play a greater role. The operative mode frequently seems to be trial and error. Large corporations and other organizations are too often rigid and slow to adapt; innovation is more likely the product of new entrants with an old-fashioned entrepreneurial spirit of risk taking. These themes have been emphasized in particular by the writings of "Austrian" economists, following in the tradition of Von Mises and Hayek who emphasize the merits of the market as an information mechanism.[11] Students of socialist planned economies contrast their rigidities with the adaptability and dynamism of western market economies. Studies of formal public planning in the United States similarly find a record of failure and lack of influence.

Proponents of selling the public lands have been strongly influenced by such thinking. Government suffers by comparison with the private sector in two critical respects. First, large government bureaucracies that try to plan formally and precisely will be defeated by the inherent unpredictability of events. Moreover, in government the general problems of large organization are complicated by the incentives of political decision making. In a world of narrow interest groups there is no constituency for efficiency in resource allocation—or, as frequently is the case, for equitable allocation. In short, scientific management in government is usually an illusion or fiction.

Such conclusions have comported well with the traditional advocacy of economists of the free market. However, it is important to recognize that current disposal proponents do not so much advocate the perfection of markets as the imperfection of political resource allocation. If government does most things poorly, especially tasks that are technically and managerially complicated, how can it be expected to manage the public lands efficiently and equitably?

Speaking personally, I find much in this critique that describes the historic record of public land management. I would put the greatest emphasis on the difficulty (or shall we say, impossibility?) of reconciling scientific management with political decision making. A rapidly changing environment in fact does pose great difficulties for effective planning and conscious design. But I am impressed with the fact that in a free competition the large corporation—essentially an embodiment of the scientific management approach—has replaced the classic liberal organization of a market of small firms. Indeed, the best prospect for achieving the scientific management sought by Pinchot and other conservationists seems to lie in the large private corporation—a conclusion not much more appealing to true free-market proponents than to proponents of a larger public sector.

The progressive conviction that scientific management can be so scientifically precise as to eliminate wide administrative discretion seems farfetched. True scientific management requires the application of critical intelligence, still leaving broad discretion for management decision making. It is this necessary discretion that is troublesome in government when the objectives for government actions are often poorly defined and political pressures are so intense. Yet, attempts to limit discretion by establishing formal rules and other constraints on government administrators often seem to produce arbitrary and inefficient results.

The case seems to me weakest for keeping the government in the business of managing production of public timber, coal, forage, and other commodity outputs—a virtual pocket of socialism maintained in the U.S. market system. There are few public values at stake and probably greater prospects of efficient management in the private sector.

It is said by pollsters that many Americans oppose parts of the Bill of Rights when they are taken out of a Constitutional context. Much of the same would probably be true of the U.S. capitalist system. The emphasis on private selfishness and competition for profits are offensive to deeply held American values of community and equality. Indeed, Christianity and capitalism have coexisted over the centuries only with severe strains. On the surface, socialist ideals of government acting for the benefit of all the people appeal much more to traditional religious values. All these concerns are heightened in an economy dominated by large private corporations rather than the small individual entrepreneurs of classic liberal thought. It is likely that American capitalism could never be accomplished today by an affirmative act, but exists only because it is a historic inheritance. Whatever the conclusions of economists, this tension between the values of the market and other American values creates a major obstacle to disposal of the public lands.

A narrower and more resolvable objection to public land disposal concerns nonmarket outputs of the public lands, for which private incentives may diverge markedly from social goals. The largest class of such outputs involves recreational use of public lands.

There are four basic reasons why the opportunities available for recreation under private landownership might be significantly less than socially appropriate. First, collection costs may be too high to justify the opening of private lands to public access on a profit-making basis. Second, some recreational use may have a very low or even zero marginal cost, but private providers would charge a higher price, preventing fully efficient use. Third, some recreation policies such as the creation of wilderness may create benefits that are realized by people who never directly enter on the land. Such "existence values" involve a true public good and cannot be captured privately. And, fourth, equity arguments can be made against the distributional consequences of private provision of recreation.

It seems clear that for many westerners the real reason for maintaining public land ownership is to ensure continued open access for recreational use. Westerners contrast their circumstances with the more limited recreational opportunities of the East. They fear that under private ownership "the forest of no trespassing signs would be as great as the one of remaining spruce, pine and fir."[12]

Employing different tenures for different classes of public land is a potential solution to the competing arguments for private production of commodities and public provision of dispersed recreation. Where lands are primarily valuable for the former use, then private tenures could be adopted, and similarly, where dispersed recreation is the most important use, the case is strongest for public ownership.

Continued public recreational use of disposed lands can also be maintained by easements or other legal mechanisms to guarantee future access. Leshy, in chapter 11, notes that such mechanisms have often not worked well in the past. However, the presence of large numbers of existing recreationists with a long history of public land use would seem to ensure the political weight to enforce any future open-access provisions.

While there is a good case for providing dispersed recreation on public lands, private ownership would facilitate more focused recreation such as resort and second-home development. The land values generated by such development can substantially exceed the values generated by existing uses of the land. While definitive studies are lacking, in my view, the unwillingness of the public land agencies to sell lands for private recreation development is one of the major inefficiencies of the current system. Even if no other public land sales take place, it would be desirable to establish a regular procedure by which land for private recreational development can be acquired.

Transfer of Public Lands to the States

The Sagebrush Rebellion started off proposing the transfer of much of the public lands to the western states. However, its leaders then shifted toward outright disposal, only later to back off from the implications of this proposal. Such vacillation reflects both ideological and interest-group tensions among proponents of major changes in land tenure.

Some have characterized the transfer of public land to the states as the worst of all possible worlds. It would keep land management in the political thicket with all the attendant obstacles to rational management. On the other hand, it would forgo the key advantages of public management at the federal level—the greater resources to employ sophisticated management methods and the wider scope for coordination. Moreover, the press spotlight and general concentration of critical attention at the national level might spur more honest and efficient administration. As has been widely noted, the state record in managing state-owned lands has not been very good.

Nevertheless, the case for transfer to the states has merits that have not been fully aired. In many respects the rights to use of public lands have already been transferred to the states; if property is really a bundle of rights, many of the public land "sticks" are already in state possession. Thus, as noted above, 50 percent of mineral-leasing revenues now go to the western states (90 percent if the Reclamation Fund is counted). Public land managers are often highly responsive to state (and local) pressures. Since 1976, public land use plans have been legally required to be "consistent" with state and local plans, as long as no violation of federal law or policy results. In such respects transfer of public lands to the states would be an incremental change that would merely bring form into closer accord with substance. The most radical change resulting from state ownership would not be greater state control over land policies but the imposition, for the first time, of land-management costs on the states. Conversely, proposals to revitalize federal land management would be the truly radical course if they meant a major shift in influence over land policy away from the states.

State ownership of public lands would more directly tie the responsibilities for land policies to those who bear the impacts of those policies. Although the states now have a major management influence, they do not pay the cost. On the other hand, national taxpayers do not bear the adverse environmental impacts that may be associated with public land policies they favor. As a rule, the separation of responsibility for benefits and for costs makes it more difficult to achieve a happy balance. A great virtue of the market is that both these elements are united in the private producer—except to the degree that externalities are present. Transfers of lands to the states in this respect would be a halfway step toward a market solution.

If the public lands were transferred to the states, it would not necessarily mean that these lands would remain in public ownership. Indeed, one of the merits of state ownership is that it would allow states to experiment with various methods of classifying land for public or private tenures. The concept of the states as laboratories for novel policy approaches is an old and honorable one in American political thought. If, as seems very likely, there is more than one defensible way to classify land for different tenures, then trial and error among the states may be the wisest course.

Suggestions that a poor past record of state management should preclude public land transfers to the states typically do not account for three key factors. First, existing state lands are often scattered sections that are remnants of old education land grants to the states. This ownership pattern largely precludes effective state management. Second, state acquisition of large public land areas would create strong new pressures for better state management. And third, the demographic and economic changes occurring in western states are creating new urban and recreationist constituencies likely to demand much improved state management.

The best case for state ownership involves lands that are suitable for public ownership and which involve mainly use by in-state residents and businesses—such as ordinary forests and rangelands whose chief use is dispersed recreation. The case for federal involvement is strong only where individual projects have impacts of wide national significance or in "critical areas" to the whole nation. Borrowing from the terminology of private land use regulation, the national parks, wildlife refuges, wilderness areas and other specially protected areas can be considered parts of a national critical area system. The remaining public lands are then more naturally the responsibility of the states.

Ultimately, like the prospects for disposal of public lands to the private sector, the outlook for transferring lands to the states may depend on matters much broader than the quality of their management. In the early years of the nation most Americans felt a much stronger sense of community loyalty toward the states and localities in which they lived. The great expansion in the role of the federal government was associated with a shift to a much stronger sense of national community. Indeed, a number of social commentators have suggested that this shift cannot be sustained in the long run and perhaps was undesirable in the first instance. The transfer of public lands to the western states themselves could be part cause and part effect of a greater future sense of identification with state instead of national communities.

Public Land Corporations

The alternative of creating public land corporations in essence seeks to revive progressive ideals by divising a new institutional setting. Recognizing that the BLM and the Forest Service have never been able to achieve professional administration independent from frequent political interference, it may be necessary to institute a more rigid separation of land management from politics. This, in fact, is the role for the public land corporation—to pose a barrier to normal politics that would allow professional administrators to manage the lands in expert fashion. Unlike existing land management, a public land corporation would be subject to the desirable discipline that the revenues must exceed the costs. Indeed, the public land corporation would function very much like a large private corporation.

In that case, why not simply sell the land to a large private corporation? One reason would be that a public corporation would have less finality; if the idea turned out to be a mistake, the public corporation could be abolished readily, unlike a private corporation. However, the public corporation would also be susceptible to future political manipulation, even if it met all its aims. The risk is that a successful public land corporation would get caught in political crossfires and then lose its independence from politics. In short, lack of permanent private status avoids some risks but creates others.

A public corporation may seem more acceptable because it more easily could be influenced to heed nonmarket values. Yet, steps in this direction will require political involvement that again may compromise the objective to achieve professional administration by experts.

The case for a public corporation is strongest in areas of the public lands where the primary use is commodity production. In particular, prime timberlands might be managed in this fashion. Public timber corporations could also provide a particularly effective transitional mechanism for shifting prime timberlands from public to private ownership. The federal government (along with other governments currently receiving shares of public timber revenues) would initially own all the stock in the new timber corporation. Gradually—over perhaps as long as ten to twenty years—the stock could then be sold off in securities markets. Such sales might be more orderly and bring a higher financial return than competitive bidding for the land itself. While small investors would be precluded from bidding for large blocks of timberlands, they could acquire shares of stock in timber corporations on an equal footing with large investors.[13]

Long-Term Leasing of Public Lands

A long-term leasing alternative is similar to the public corporation in that it is a way of avoiding any irreversible commitments. The lease will expire at some point, and the government could then decide not to renew it. To be sure, the objective to turn management over to the private lessee requires a security of tenure inconsistent with frequent review of lessee status. Yet, the more distant the lease renewal and the more secure the tenure, the closer the practical consequence comes to outright disposal. The differences between a 100-year lease and sale of the land are largely symbolic.

Mineral leases now last as long as the mineral production continues. However, they are renegotiated at periodic intervals, which can create significant uncertainties about the future. More important, most mineral leases contain requirements for diligent development. The effect of this requirement is to eliminate private incentives to conserve public resources. The government itself therefore must decide when a particular mineral deposit should be produced, often a difficult matter to determine. For mineral leases, extension or abolition of the diligent development requirement would do more than a longer lease term to promote private assumption of responsibility for resource management.[14]

Probably the best case for long-term leasing can be made for public grazing lands. The productivity of many grazing lands is so low that significant public management expenses cannot be justified. At present public management costs far exceed any revenues received by the government. The ranchers who use the lands know them well and are well equipped to make investment and management decisions. If ranchers are provided a secure long-term tenure, they will have the necessary incentive to maintain the lands in good condition. Indeed, their stake would be greater than that of individual public agency personnel temporarily assigned to manage the lands. Long-term grazing leases would also have to include performance standards for the achievement of non-grazing outputs on the lands. It would also be important to provide a readily available mechanism for abandoning grazing and conversion to higher values uses that might emerge. For example, it is already apparent in some areas that recreational "ranchettes" are creating rangeland values far higher than livestock grazing can justify.

Leasing is in fact a very broad and flexible concept, which can range from very tight public control over lessee use of the land to almost total lessee discretion. As a practical matter, the former does not differ so much from direct government production of the resource or the latter from outright disposal.

Leasing may offer some advantages over disposal in protecting dispersed recreational use of the land. While a lease stipulation may be technically identical to a deed easement, enforcement of the stipulation may be facilitated by the less-proprietary attitudes associated with leasing. In many rangeland areas the recreational use of the land is considerably more valuable than the grazing use, despite the fact that recreation receives less management attention. It would be important not to jeopardize this recreational use by any sale of the land.

The Clawson Leasing Proposal

The Clawson paper contains a detailed proposal for a long-term leasing system. I agree with Clawson that a long-term leasing system probably has a greater prospect of being adopted than any other form of major tenure change. It could accomplish the objective of turning more management responsibility over to the private sector with less likelihood of raising the fierce opposition that outright disposal might provoke. While applauding the general concept, I do have some questions and concerns about some of the specifics of Clawson's leasing proposal.

The leasing of timberlands faces significantly greater obstacles than rangelands. Even a 100-year lease may not always be long enough to cover two harvest rotations. In any case, the end of any given lease term is unlikely to coincide precisely with the proper time to harvest the timber, raising the issue of the incentive to reforest. It seems to me that a more secure incentive for reforesting and other forest investments is needed than the prospect of receiving the value of the standing timber at the end of the lease term. Estimating this value will be subject to many uncertainties; lessees are likely to be deterred by the absence of a well defined payment. Generally, timber leases would have to have such a long term that the land might just as well be disposed outright. On the other hand, the Clawson suggestion to lease public lands to recreation groups for private recreational provision seems to me especially promising.

The pullback provisions proposed by Clawson have several key problems:

- Acquirers of the initial lease will face substantial uncertainty because they will not know just what land will eventually be pulled back by another lessee. Some potential users may require all the land or want none of it; removing up to one-third may render the remaining two-thirds of minimal value.
- Provisions for subleasing need to be carefully considered. Circumstances toward the middle and later years of a fifty-year lease term may be greatly altered. Adjustments in the size of leases or provisions for new uses of leased lands may be necessary. Such adjustments would be greatly facilitated by subleasing. Subleasing would provide a market mechanism for determining desirable changes from one use of land to another. In short, the pullback provision is a substitute for subleasing, but a rigid one and one not available frequently enough.
- The pullback provision could artificially stimulate excessive intermingling of uses. For example, commodity leases may be for 25,000 acres or less, yet satisfactory provision for a wilderness area require a large contiguous block of land of 200,000 acres. Pullbacks for wilderness thus might yield a scattered set of wilderness parcels less suited to wilderness management. The pullback proposal generally does not seem to recognize the necessity in some cases for large contiguous blocks of land in the same use.
- The failings noted above in the private market provision of dispersed recreation may result in much less than the socially optimal level of recreational leasing. For example, to the extent that a large part of wilderness value consists of a national existence value, the pullback approach (or any private market approach) will not provide for sufficient wilderness. Hence, long-term leasing—with or without a pullback provision—will still require specific allocation by the government of some land for uses such as wilderness.

In general, the pullback approach would introduce a complex set of strategic calculations for public land users. It is difficult to project just how these calculations would work out. Given such uncertainties, I need more persuasion on the merits of the pullback approach.

Conclusion

Clawson concludes his chapter on an upbeat note with regard to the prospects for improving the management of the public lands, whether by significant tenure changes or other means. Leshy, on the other hand, doubts that a coalition of interest groups can be forged that will support any major changes.

I suspect, however, that ideology may play a much greater role in the future than Leshy's view suggests. Indeed, the environmental movement of the 1970s showed the power of ideas as well as that of interest groups. The creation of the national wilderness system is partly testimony to the great appeal of the idea of wilderness to millions of Americans—many of whom will never visit a wilderness area. The protection of endangered species, involving few direct beneficiaries (human that is), and having no immediate constituency, still commands wide political support as an important goal of public policy.

Leading political scientists have been reassessing the emphasis put on interest groups in recent theories of the political process. Harvard professor James Q. Wilson not long ago commented on "the importance for policy making of the ideas of political elites. A political scientist such as myself, trained in the 1950s when politics was seen almost entirely in terms of competing interests, was slow to recognize the change."[15] It is now thirty years since the publication of perhaps the most influential statement of interest-group liberalism, *The Governmental Process* by David Truman.[16] Keynes once suggested that men of affairs are fated to follow the defunct theories of leading academicians of many years past. Perhaps, having been warned, we should make a special effort to avoid this fate.

If the politics of the future is as much or more a competition of ideas and ideologies as of interest groups, then a key role will be played by intellectuals and other arbiters in ideological matters. The press and other communications media will have a critical role, a development already much noted. Workshops such as this one which seek to rethink the fundamental issues of public land ownership may have a greater impact than is commonly supposed. If an intellectual consensus were to develop for major changes in public land tenure, I have the feeling that the political process would eventually follow.

It is obviously highly speculative to suggest where such a consensus might lie. However, I will venture a few thoughts that are already reflected in earlier remarks.

It is hard to imagine any brand new ideology emerging which seems definitively to resolve the long-standing questions of the roles of democratic politics, professional expertise, the judiciary, the private sector, and so forth. The current age seems particularly unsympathetic to absolute or utopian solutions. Rather, each ideology is perceived as having its valid points and less valid points—its benefits and costs as it were.

296 PART IV: INTERMEDIATE POSITIONS AND SPECIAL PROBLEMS

The objective in reevaluating land management may be to match up different ideological approaches to the specific circumstances of specific types of land. In short, as suggested previously, different areas of public land may have to be classified for different tenures that reflect suitable governing and managerial philosophies.[17]

Thus, lands with few commodity uses and extensive dispersed recreation do not involve highly complex management decisions. The gains from more scientific and sophisticated management of these lands would be much less than in other areas; hence, the costs of the democratic politics of interest-group liberalism (that is, multiple-use management) are not so great. Relative to commodity production, the private market also fares poorly in providing recreation. In sum, lands with the characteristics above probably should be maintained in public (if state or local) ownership.

High-value timberlands in lowland areas with only modest recreational use provide an opposite circumstance. The costs of unsophisticated and inefficient public management of the timber are high, while the losses to recreation and other nonmarket values resulting from private management would be relatively much less than in other areas. In fact, private timber companies currently often encourage recreational use of their lands. Private ownership and management of commercial timberlands thus may well be indicated. Public rangelands provide an intermediate case, often involving low commodity combined with low or modest recreation values.

In some ways this classification strategy for public land tenure resembles interest-group liberalism in its approach. However, rather than a compromise among the narrow immediate interests of particular groups, it may more importantly represent a viable compromise among groups espousing broad ideologies. It also has elements of progressivism in that it would seek to draw a line beyond which expert management could operate without frequent political interference. However, the instruments for achieving such expert management would be in the private sector, in many cases large private corporations, rather than government agencies.

In conclusion, it should be very apparent by now that the current rethinking of the federal lands illustrates much broader issues facing American government today. The future of these lands raises questions of the magnitude of socialism versus capitalism, or nationalism versus states rights. Such matters arise in a particularly sharp form because of the anomalous character of public land ownership. In a nation supposedly committed to private ownership, more than one-third of the land is public. In a nation supposedly committed to decentralized authority, public land management is centralized at the national level. To be sure, there may be good reasons why the federal lands constitute valid exceptions. Another possibility is that American commitments to private and decentralized institutions today have become more mythical than real. A third important possibility is that federal land ownership is mostly an accident of history, and that major changes would follow from an effort to make land tenure accord with basic economic and political principles.

Notes

1. See Theodore J. Lowi, *The End of Liberalism: Ideology, Policy and the Crisis of Public Authority* (New York, Norton, 1969).

2. Samuel P. Hays, *Conservation and the Gospel of Efficiency: The Progressive Conservation Movement—1890–1920* (Cambridge, Mass., Harvard University Press, 1959).

3. *Ibid.*, pp. 168–269.

4. Herbert Simon, *Administrative Behavior* (New York, Macmillan, 1947); David B. Truman, *The Governmental Process* (New York, Alfred A. Knopf, 1951); and Dwight Waldo, *The Administrative State; A Study of the Political Theory of American Public Administration* (New York, Ronald Press, 1948).

5. John Kenneth Galbraith, *American Capitalism: The Concept of Countervailing Power* (Boston, Mass., Houghton Mifflin, 1956).

6. Marion Clawson, "The National Forests," *Science*, Feb. 20, 1976, p. 766. Also see Marion Clawson, *The Economics of National Forest Management* (Washington, D.C., Resources for the Future, June 1976).

7. John Baden and Richard Stroup, "Saving the Wilderness," *Reason* (July) 1981.

8. See John V. Krutilla and John A. Haigh, "An Integrated Approach to National Forest Management," *Environmental Law* (Winter) 1978.

9. See Christopher K. Leman and Robert H. Nelson, "Ten Commandments for Policy Economists," *Journal of Policy Analysis and Management* (Fall) 1981.

10. Charles Lindblom, "The Science of Muddling Through," *Public Administration Review* (Spring) 1959.

11. Friedrich A. Hayek, *Individualism and Economic Order* (Chicago, Ill., University of Chicago Press, 1948).

12. Dan Abrams, "The Rebellion is Getting Hot," *Jackson Hole News*, Dec. 5, 1979.

13. See Robert H. Nelson, "Making Sense of the Sagebrush Rebellion: A Long Term Strategy for the Public Lands," Paper presented at the Third Annual Conference of the Association for Public Policy Analysis and Management, Washington, D.C., October 23–25, 1981. A shortened version of this paper is contained in Robert H. Nelson, "A Long Term Strategy for the Public lands," in Richard Ganzel, ed., *Resource Conflicts in the West* (Reno, Nevada Public Affairs Institute–University of Nevada, 1983).

14. See Robert H. Nelson, *The Making of Federal Coal Policy* (Durham, N.C., Duke University Press, 1983). Also Robert H. Nelson, "Undue Diligence: The Mine-It-Or-Lose-It Rule for Federal Coal," *Regulation* (Jan.–Feb., 1983).

15. James Q. Wilson, "What Can Be Done?," in *AEI Public Policy Papers* (Washington, D.C., American Enterprise Institute, 1981) p. 19.

16. Truman, *The Governmental Process*.

17. See also Robert H. Nelson, "The Public Lands," in Paul R. Portney, ed., *Current Issues in Natural Resource Policy* (Washington, D.C., Resources for the Future, 1982).

Index

Acquired Lands Mineral Leasing Act
  (1947), 256
Act of 1891, 49–50
Agricultural College Act (1862), 40
Alaska
  public lands in, 77, 79–80
  wolf extermination in, 244
Alaska National Interest Lands Conserva-
  tion Act (1980), 77, 80, 83
Alaska Native Claims Settlement Act, 76
Alaska Statehood Act, 76
Andrus, Cecil, 120
Aspinall, Wayne, 54
Asset management. *See* Privatization.
Autonomy, 133–134, 136

Baden, John, 166, 171
Barnett, Richard, 152
Big business, 46–47
Big Sky development, 150–151
Boulding, Kenneth, 190–191
Bromley, Daniel, 176
Bruner, William, 166
Bureaucracy
  and appropriations maximization, 189
  incentives for, 285
  performance of, 11, 129
  and self-interest, 22, 161

Bureau of Land Management (BLM), 64
  Alaskan role of, 80
  forestland of, 84
  and grazing rights, 86–87
  and land disposal, 9
  leasing programs of, 78, 213, 252–253
  management costs of, 280
  organic act for, 54
  and politics, 291
  range improvement by, 167
  timber management of, 85
  unneeded land of, 213
Bureau of Reclamation, 42, 45, 79

Canada, 27
Capitalism, 287
Capture, 23, 177, 280
Carter administration, 130
Chaining, 167
Clarke–McNary Act, 53
Clawson, Marion
  as critic, 280
  on Forest Service, 168
  leasing proposal of, 293–294
Clay, Henry, 37
Coal
  development of, 243

Coal (*continued*)
  from federal lands, 89, 245, 250
  severance taxes on, 268
Collective
  and behavior, 8, 138
  as environmental interest, 155
  failure of, 22
  preferences, 20–21, 131–147
  shaping power of, 135–136
  state action, 138–139
  values of, 139–140, 143–144
Collectivities
  as administratively efficient, 134
  entry and exit, 136–137
  mechanisms of, 134–135
Colorado Board of Land Commissioners, 29
Colorado River basin, 262
Community values, 133, 140
Compensation, 257
Conservation
  and economics, 283
  leasing for, 227–228
  organizations, 212
  progressive, 85
Cornell University, 41
Corporations, as landowners, 46, 154
Craig, George, 30
Culhane, Paul, 177

DeAlessi, Louis, 176–177
Deregulation, 275
Desert Land Act, 49
Disposal. *See* Privatization.
Distributional equity, 158, 159
Divestiture. *See* Privatization.
Dix, Dorothea, 40
Dowdle, Barney, 167

Easterners, 14
Economic efficiency. *See also* Externalities.
  defined, 158
  in investment, 165
  and multiple use, 163
  norms of, 159
  and public ownership, 129
Economics, 285–289
Economic system, 67–69
Edwards, Frank, 29

Efficiency. *See* Economic efficiency.
Electorate, 151
Enclosure Acts, 187, 188
Energy Research and Development Administration, 79
Energy resources, 115
Enlarged Homestead Act, 51
Environment
  and market system, 4
  movement for, 145
  and public land, 103–104, 153
Equity. *See* Distributional equity.
Evans, John V., 55
Executive Order 12348, 157
Existence values, 174, 288
Experts, 284
Externalities, 183. *See also* Economic efficiency.
  importance of, 5–6
  and land tenure, 16
  and privatization, 175–176
  and public management, 11

Federal debt, and land sales, 127–128, 171–172, 214

Federal government, 1–2. *See also* Government.
  constitutional restraints on, 236–237
  as proprietor, 112
  restrictions of, 249–254
  as sovereign, 111–112
Federal Highway Act, 267
Federalism, 266
Federal Land Policy and Management Act of 1976 (FLPMA), 2, 20, 54–55, 75, 98, 109, 116–118, 163, 165, 195, 202, 203, 279
  and environmental concern, 126
  and land use planning, 119–120
  process of, 120–121
  and retention, 126
  and wildlife management, 244
Federal lands. *See* Public lands.
Fernow, Bernhard, 51, 71
FLPMA. *See* Federal Land Policy and Management Act of 1976.
Forest and Rangelands Renewable Resources Planning Act (1974), 98–99, 109, 163, 195, 202, 203, 279

Forestland
    changing role of, 184–187
    federal purchase of, 53
    and interest rates, 186
    for lease, 224–225
    ownership of, 44
Forest Reserve Act, 49–50
Forestry
    economics of, 187–90
    professional, 51
    schools of, 63
Forest Service, 52, 64. *See also* National
    forests.
    acquisitions of, 53
    commercial forests of, 217
    inefficiency of, 167–168, 283
    lands managed by, 75
    management costs of, 280
    planning process of, 99
    and politics, 291
    research program of, 78
    timber management of, 85, 253
    unneeded land of, 213
Forest Service Organic Act (1897), 2, 52
Free land, 69–70
Free-rider problem, 140, 173
Frontier, 212

General Land Office, 43
George, Henry, 45
Government. *See also* Federal govern-
    ment; Local government; State
    government.
    and collective preferences, 131–132
    as commons, 162
    and competing values, 142
    corruption of, '142
    and economic efficiency, 6, 68, 154,
        176
    incentive structure in, 161
    and individual preferences, 131
    and interest groups, 142–143
    as land use regulator, 129–130
    objectives of, 21
    and political markets, 161
    scientific administration of, 277
Governmental Process, The (Truman),
    295
Gray, Lewis C., 45

Grazing, 86–88, 115, 246, 250
    allocation regulations, 168–169
    and equity, 153
    lease of, 226–227, 292
    permits for, 96–97, 214
    and range improvements, 167, 268
    and sustained yield, 164
Greeley, Horace, 42
Groundwater, 260

Hagenstein, Perry, 166
Hamilton, Alexander, 36
Hanke, Steve, 125
Hatch Act (1887), 38
Hibbard, Benjamin, 51
Historic preservation, 146
Homestead Act (1862), 68–69
    and speculators, 42–43
Homestead movement, 42, 50–51
Hooker Chemical Company, 153
Hoover, Herbert, 270
Housley, Ray, 29
Hyde, William, 167, 168

Ickes, Harold, 45
Ideology, 4, 28, 275, 295–296
Incrementalism, 285
Indian reservations
    overgrazing on, 167
    water for, 259, 264
Individual autonomy, 133–134, 136
Inheritance, 145
Interest-group liberalism, 27–28, 275
    challenge to, 279–282
    and progressivism, 276–279
    and property rights, 282
Interest groups, 129, 150, 154. *See also*
    Capture.
    and government, 142–143
    and public lands competition, 162
    and scientific management, 97, 284,
        286–287
    and use restrictions, 252
Investment, 165, 171
Irrigation, 258

Jefferson, Thomas, 7, 36
Johnson, Ronald, 167

Kremp, Sabine, 167

Land
  prices of, 149–150, 172
  use planning, 98–100
  values, 153
Land grants
  income from, 40
  to railroads, 43–44, 47, 241, 242–243,
    249, 252
  for schools, 39, 206–207
  for social programs, 38–39
  to soldiers, 38
  to states, 206–210, 240–241, 249
  for transportation, 39
Land Ordinance of 1785, 37, 39, 47
Lane, Franklin K., 52
Leasing, 25, 28, 30, 219–222
  basic goal of, 221–222
  and collusive action, 230–231
  and private management, 14–15
  proposal for, 222–229
  pullback, 229–232, 294
Libecap, Gary, 167
Live oak, 47
Local government
  and federal lands, 113
  impacts on, 265–268
  payments to, 100–103
Love Canal, 152–153
Lumber industry, 70

Mann, Thomas, 133
Market
  belief in, 4–6, 190
  failure, 10, 62–64, 154–155, 176
  as information mechanism, 286
  and long-term values, 119
Mineral King ski area, 248
Mineral Leasing Act (1920), 245, 255, 257
Minerals, 89
  and damage compensation, 256–257
  on federal lands, 88–90, 115, 254–257
  leases for, 224, 245–246, 292
  policy, 47
  rights to, 15, 76, 241, 256
Mineral springs (Arkansas), 48
Mining Law (1872)
  and mineral rights, 88
  replacement of, 255
Morrill, Justin Smith, 40–42
Morrill Acts, 38, 41

Morse, Harold, 152
Muckrakers, 46
Multiple use, 3, 116–117, 271–272
  as balance, 279
  financial implications of, 269
  land in, 19, 82–83
  as management goal, 163
  and ownership patterns, 170, 183
  priorities of, 92–93
  receipts from, 94 (table)
  and recreation, 91
  and restrictions, 249
Multiple Use and Sustained Yield Act
  (1960), 203, 279

National Environmental Policy Act
  (NEPA), 19, 98
  and Environmental Impact Statements,
    99, 103
National Forest Management Act (1976),
  109, 163, 195, 202, 279
National forests, 77–78. See also Forest
  Service.
  in Alaska, 80
  commercial forestland on, 84–85
  public recognition of, 109
  timber sold from, 85–86
  underharvesting of, 166
National Park Service
  and concession contracts, 253
  management purposes of, 78
National Park System, 48–49, 78
National Wilderness Areas Preservation
  System, 81
National Wildlife Refuge System, 78
Natural gas, 89
Nelson, Gaylord, 30
Nelson, Robert, 216–217
NEPA. See National Environment Policy
  Act.
Newlands Reclamation Act, 42
Noh, Laird, 29
Northwest Ordinance (1787), 35, 37

Oil, 89
One-Third of the Nation's Land, 54
Our Land and Land Policy (George), 45
Outdoor recreation. See Recreation.
Outer Continental Shelf (OCS), 78–79
  leasing of, 213

oil receipts from, 269
seafloor of, 76
Ownership. *See also* Private ownership;
    Public lands; State ownership.
    debate over, 130
    and federal land management, 128–129
    and regulation, 131–132, 146

Payments-in-lieu-of-taxes, 101–103
Payments in Lieu of Taxes Act (1976),
    267
Pinchot, Gifford, 51–52, 71, 110, 277
Planning, 202–203
PLLRC. *See* Public Land Law Review
    Commission.
Powell, John Wesley, 278
Preemption Law and Timber Culture
    Act, 42–43
Preference, 132–133
Preservation, 140, 150
Price
    in allocation processes, 160–161
    as information, 170
    and resources, 152
Private ownership
    arguments for, 12, 15–16
    benefits from, 170–171
    and commodity production, 187
    and economic efficiency, 23, 182–183
    and environmental quality, 153
    and federal mineral rights, 88
    as ideology, 4
    and public goods, 174
    and wealth maximization, 160
Privatization, 8–9, 65–66, 67, 125,
        129, 169–173, 181, 210–214. *See also*
        Public lands; Retention; State owner-
        ship.
    and allocation decisions, 160
    and Asset Management Program, 8–9,
        149
    criticisms of, 30, 173–177
    and current users, 172
    definitions of, 109–110
    and disposal program, 171–173
    and equity, 23, 127
    gains from, 10, 178, 188–189
    and land values, 127
    patterns of, 238–239

and protective covenants, 26, 104,
        237–238, 239–240, 253
    and regulation, 131–132, 147
    by sale, 285–289
    social savings from, 188
    and timber, 189–190
    and Western subsidization, 270–271
Progress and Poverty (George), 45
Progressivism, 2, 18
    and conservationism, 275
    and interest-group liberalism, 276–279
Property Review Board, 157
Property rights
    controls on, 146
    and factor price shifts, 23–24
    and resource utilization, 182
    rules of, 145
    and social values, 147, 181–182
Property taxes, 111
Prospectors rights, 103
Public, types of, 110–111
Public choice theory, 182
Public corporations, 25, 28, 214–219
Public debt, 127–128, 171–172, 214
Public Debt Act (1790), 37
Public domain. *See* Public lands.
Public goods, 174, 183
    and collective action, 6–7
    from forestland, 63
    importance of, 5–6
    and land retention, 16
    and public management, 11
Public Land Law Review Commission
        (PLLRC), 2, 19, 198, 241
    establishment of, 54, 110
    on land use planning, 98–99
    on mineral rights, 88–89
    on pricing, 95
    on retention, 105
    and tax immunity, 101
Public lands, 76 (table), 76–83, 162. *See*
        also Privatization; Public manage-
        ment.
    acquisitions, 53–54
    allocation of, 158–164
    capital improvement fund, 119
    capital value of, 198, 201
    classification strategy for, 296
    corporations for, 291–292
    dedicated, 80–81

Public lands (*continued*)
  economic contribution to, 19
  and economic efficiency, 10–13, 112,
      281
  and equity, 153
  federal–state issues, 120, 236–237
  future of, 199–222
  history, 17–18, 65–67
  and intellectual history, 27
  investment policies, 164–169
  leases of, 89, 220, 252–253, 292–293
  as leftovers, 92
  as limited resource, 104–105
  and local tax bases, 29
  management of, 3, 13–17, 24, 96–100,
      114–128, 157, 160–166, 235–236,
      277–278, 280, 282–285
  in new states, 35–36
  noncommodity uses on, 83
  original rationale for, 2–3
  permanent ownership, 126
  and political process, 7
  private benefits from, 268
  productivity of, 92
  publics of, 110–113
  reacquisition of, 240
  for recreation, 47–48
  restrictive arrangements, 105
  revenue from, 37–38, 40, 55, 93–96
  revenue sharing from, 101, 267–268
  role of, 196–199
  and secondary uses, 237–249
  and social philosophy, 7
  states, 206–207
  status changes of, 75–76
  theft of, 66
  transfer to states, 289–291
  unsurveyed, 43
  uses of, 83–96, 128
  values of, 12–13, 69, 115, 144–147, 212,
      287–288, 291
  and water, 258, 260–263, 264
Public management
  arguments for, 12, 184
  capital account for, 203–204
  dissatisfaction with, 195–196
  efficiency of, 11, 22–23, 166–169
  objectives of, 189
  and private use, 197
  revolving funds for, 205

  and social optimum, 11–12
  system of, 218
Public ownership
  and efficiency, 176, 178
  and interest rate, 186
  and multiple use, 183
  patriotic arguments for, 183–184
  symbolic significance of, 20
Public Rangelands Improvement Act
    (1978), 165, 268
Pullback. *See* Leasing.

Reagan, Ronald, 118
  and privatization, 125–127, 157, 213
  and states' rights, 156
Reclamation Fund, 281
Recreation, 90–91, 247
  and commercial development, 16, 248
  fees for, 95, 204, 281
  lands for, 14, 81, 115, 216
  and leasing, 221, 224, 293
  markets for, 175
  and preservation emphasis, 150
  and private ownership, 174, 281
  values, 186–187
Recreation and Public Purposes Act, 241
Reforestation, 185
Renewable Resources Planning Act
    (1974). *See* Forest and Rangelands
    Renewable Resources Planning Act
    (1974).
Rents, 154, 188
Resettlement Administration, 45
Retention, 2. *See also* Privatization;
    Public lands.
  and allocation decisions, 160
  and conservationist ideology, 277
  defense of, 276
  and management, 200–206
  original rationale for, 18
  and public values, 21
Revenue sharing. *See* Public lands.
Revolutionary War, 36
Robber barons, 46
Robinson, Joan, 190
Rogers, Rowena, 29
Roosevelt, Theodore, 2, 278
RPA. *See* Forest and Rangelands Renew-
    able Resources Planning Act (1974).

Sagebrush Rebellion, 1, 29, 55, 109, 118, 157, 196, 266, 270
Schurz, Carl, 51, 71
Scientific management, 3, 18
  as illusion, 28
  and interest groups, 286–287
  and private corporations, 287
Silviculture, 185
Simon, Herbert, 278
Small Tract Act, 243
Smith, Adam, 66, 182
Socialism, 65, 68
Social utility, 190
Sowell, Thomas, 160
Speculators, 152
Squatters, 37
State
  and collective values, 21, 144
  goals of, 141–142
  majoritarian acts of, 141
  as political collective, 138
State government, 265–268. See also Government.
  federal lands jurisdiction of, 113
  land claims of, 35
  leasing by, 228
  management by, 24, 210, 290
  payments to, 100–103
  as policy laboratories, 290
  and severance taxes, 257
  and water resources, 259
State ownership, 28, 156. See also Privatization; Public lands.
Stock Raising Homestead Act (1916), 51, 52, 60, 60n, 241
  and mineral rights, 243, 250, 256
Stroup, Richard, 166, 171
Surface Mining Control and Reclamation Act (1977), 88–89, 239, 243
Sustained yield, 163–164
Sustained Yield Forest Management Act (1944), 253

Taxpayers, 162
Taylor, Edward T., 52, 270
Taylor Grazing Act (1934), 2, 52, 203
  and base property, 246
  and homesteading, 278
  as policy change, 53
  and privatization, 211–212

Technology, 182
Teeguarden, Dennis, 217–218
Timber, 84–86, 115
  cut-and-run exploitation, 70–71
  dual production system, 63
  and government management, 62–63, 167–168, 246–247, 281
  industry, 30, 166
  prices of, 95–96, 185
  private value of, 96
  and pullback, 229–230
  supply of, 49, 78
  sustained-yield rule, 87
  and technological change, 185–186
  and water rights, 263
Timber and Stone Act, 50
Timber Culture and Preemption Acts, 49
Transaction costs, 164
Trans-Alaska oil pipeline, 248–249
Tree farming, 71, 184–185
Truman, David, 278, 295
Turner, Frederick Jackson, 61

United States Congress, public lands authority of, 266
United States Department of Agriculture—Forest Service. See Forest Service.
U.S. Department of Defense, 79
U.S. Department of the Interior, 265
Users, 113–114
  expectations of, 128
  fees for, 165

Values
  collective versus individualistic, 130–131
  and public lands, 201
Voluntarism, 136
Voters, 151–152

Waldo, Dwight, 278
Walker, Thomas B., 44
Water
  allocations of, 263–264
  competing interests, 261
  and land tenure, 14
  and privatization, 26
  and public lands, 175, 258–265
  rights to, 243–244

Watershed protection, 91–92, 260–262
Watt, James, 118, 120, 156
Weeks Act (1911), 2, 18, 53, 240, 260
Western states
  and federal lands, 26
  payments to, 102 (table)
  socializing of, 64–65
Weyerhaeuser, Frederick, 44
Wilderness, 3, 81
  idea of, 295
  and leasing, 224, 226, 251
  opportunity cost of, 5
  and private ownership, 174
  and pullback, 229–230

Wilderness Act, 250–251
Wildlife, 91–92, 115
  and privatization, 175–176
  state management of, 244
Will, George, 270
Wilson, James Q., 295
Wilson, Woodrow, 277
Winters doctrine, 258

Yellowstone River, 48
Yosemite National Park, 241–242
Yosemite Valley, 48

Zoning, 130